Histories of Productivity

Global issues such as climate change and the aftermath of the 2008 financial crisis have spurred interest in thinking about the history of the modern economy that goes beyond disciplinary economic history. This book contributes to the cultural history of capitalism and its different regimes of productivity by pursuing the perspective of body history and by providing a global scope. Throughout modernity, the body served as a fundamental, albeit essentially changing, linchpin for both the organization of economic practices and for intellectual reflections on the economy. In particular, it was the pivotal interface to render notions of economic productivity intelligible. The book explores this central thesis in a range of case studies, drawing on source material from West Africa, Europe, Mexico, and the US. Framed by a theoretically informed introduction, which also provides a conceptual history of notions of productivity, and by an afterword that brings the approaches explored in this volume into dialogue with scholarship inspired by Marx and Foucault, the individual chapters tackle the concept of productivity from a wide array of angles, each illuminating the promises and problems of a cultural take on the history of economic productivity.

Peter-Paul Bänziger is currently the recipient of an Ambizione Grant from the Swiss National Science Foundation at the University of Basel.

Mischa Suter is a lecturer in Modern European History at the University of Basel.

Routledge Studies in Modern History

For a full list of titles in this series, please visit www.routledge.com

13 **War and Displacement in the Twentieth Century**
Global conflicts
Edited by Sandra Barkhof and Angela K. Smith

14 **Longue Durée of the Far-Right**
An international historical sociology
Edited by Richard Saul, Alexander Anievas, Neil Davidson and Adam Fabry

15 **Transnational Perspectives on Modern Irish History**
Edited by Niall Whelehan

16 **Ireland in the World**
Comparative, Transnational, and Personal Perspectives
Edited by Angela McCarthy

17 **The Global History of the Balfour Declaration**
Declared Nation
Maryanne A. Rhett

18 **Colonial Soldiers in Europe, 1914–1945**
"Aliens in Uniform" in Wartime Societies
Edited by Eric Storm and Ali Al Tuma

19 **Immigration Policy from 1970 to the Present**
Rachel Stevens

20 **Public Goods versus Economic Interests**
Global Perspectives on the History of Squatting
Edited by Freia Anders and Alexander Sedlmaier

21 **Histories of Productivity**
Genealogical Perspectives on the Body and Modern Economy
Edited by Peter-Paul Bänziger and Mischa Suter

Histories of Productivity

Genealogical Perspectives on the Body and Modern Economy

Edited by Peter-Paul Bänziger and Mischa Suter

Routledge
Taylor & Francis Group
NEW YORK AND LONDON

First published 2017
by Routledge
711 Third Avenue, New York, NY 10017

and by Routledge
2 Park Square, Milton Park, Abingdon, Oxon OX14 4RN

First issued in paperback 2018

Routledge is an imprint of the Taylor & Francis Group, an informa business

Library of Congress Cataloging-in-Publication Data
Names: Bänziger, Peter-Paul, 1977– editor. | Suter, Mischa, editor.
Title: Histories of productivity : genealogical perspectives on the body and modern economy / edited by Peter-Paul Bänziger and Mischa Suter.
Description: New York : Routledge, 2017. | Series: Routledge studies in modern history ; 21 | Includes bibliographical references and index.
Identifiers: LCCN 2016013716 (print) | LCCN 2016022802 (ebook) | ISBN 9781138696983 (hbk : alk. paper) | ISBN 9781315522777 ()
Subjects: LCSH: Labor productivity—Social aspects. | Production (Economic theory)—Social aspects. | Human body—Social aspects. | Economics—Sociological aspects. | Economic history.
Classification: LCC HC79.L3 H57 2017 (print) | LCC HC79.L3 (ebook) | DDC 338/.06409—dc23
LC record available at https://lccn.loc.gov/2016013716

ISBN 13: 978-1-138-32951-5 (pbk)
ISBN 13: 978-1-138-69698-3 (hbk)

Typeset in Sabon
by Apex CoVantage, LLC

Contents

Acknowledgments

The composition and production of this volume would not have been possible without the intellectual and practical support from many people. First of all, we are grateful to Marcel Streng for his conceptual contributions and for preparing parts of the introduction. Together with Lea Bühlmann and Peter-Paul Bänziger, he was also a co-organizer of the conference "Produktive Körper. Aktuelle Forschungen zur Körpergeschichte des Ökonomischen" presented through the Department of History at the University of Basel and organized in collaboration with the Arbeitskreis für Körpergeschichte. We thank Marcel and Lea for making those inspiring days in June 2014 happen; they provided the opportunity to discuss some of the basic ideas of this volume with the participants. We are also grateful to Jennifer Burri and Yves Hänggi for composing the conference minutes. The Arbeitskreis für Körpergeschichte further supported the preparation of this volume. Our special thanks go to its members, as well as to the reviewers of the proposed chapters. Jo Corrall, Bernard Heise, Kate Sturge, weiss traductions, and Christina Wheeler translated parts of this volume, edited, and proofread the proposal and the final manuscript, respectively. Funding for these services was provided by an Ambizione Grant of the Swiss National Science Foundation, by the Department of History at the University of Basel, and by the Berta Setzer Foundation. Heinrich Hartmann and Eveline Müller contributed further support, as did Jennifer Morrow and Max Novick from Routledge. Last, but not least, we would like to thank the authors for the excellent collaboration. Our very special thanks, however, go to Arina Heussler. Next to providing immeasurable intellectual support, she has done a superb job with the larger part of the administrative efforts with respect to the conference, and she masterly shepherded the book through its production.

Peter-Paul Bänziger and Mischa Suter
Basel and Zurich,
February 2016

1 Histories of Productivity

An Introduction

Peter-Paul Bänziger, Marcel Streng, and Mischa Suter

Today, productivity is a basic concept of economic science (see, for example, Field 2008; OECD 2008). Productivity indices are established and commonly used in accounting techniques—particularly for measuring input/output relations—both on the microlevel of firms, or single "productive factors" such as labor, and on the macrolevel of large-scale economic systems. First introduced by the US Bureau of Labor Statistics in the 1920s and generalized by the Organisation for Economic Co-operation and Development (OECD) during the *trente glorieuses* after World War II, the calculation of "productivity ratios" today serves as a powerful epistemological tool for the surveillance and management of economic processes in business administration, auditing, and consulting all over the world. As ecological critics have suggested since the 1970s (most influentially, Meadows et al. 1972), productivity has, together with the related concept of growth, become an end in itself and the central purpose of current economic policies. At the same time, the rise of such criticism, with its increasingly palpable impact on national and even international economic policies, indicates that the centrality of productivity within the economic discourse is being queried today.

From a historiographical perspective, however, this criticism results in being merely the latest chapter in a debate that has accompanied conceptualizations of productivity ever since the notion emerged in the eighteenth century. In the German-speaking setting, for instance, it is common for the historiography of modern economic thought to refer to the "Produktivitätsstreit" (controversy on productivity) among economists. Around 1900, many political economists in Germany, France, and, to a lesser extent, Britain attested and deplored the "vagueness" of the term. Hence, productivity may be what Walter B. Gallie (1956, 169) has called an "essentially contested concept"—one of those concepts "the proper use of which inevitably involves endless disputes about their proper uses on the part of their users". It is precisely because of its constant contestation that productivity has become one of the central themes forming the history of the modern economy—alongside growth, scarcity, labor and work, market, competition, profit, and capital, to name only some of the most important additional concepts.

It is therefore hardly surprising that debates within the discipline of history also show very divergent perspectives regarding productivity and, accordingly, many different proposals for periodization. Some historians have pointed out that "growth" as a native category is of rather recent provenance. A specific growth paradigm, they argue, can only be identified for a short span of time, from the mid-1950s to the oil crisis, that was embedded in a barely longer phase of what Charles Maier (1977) called the "politics of productivity" (see Mitchell 2011; Schmelzer 2015). Other scholars associate productivity with the mass utopias of the "short" twentieth century. According to this view, productivity—regarded as the means of bringing about affluence, mass consumption, and general happiness in an industrial society—was what fueled the defining collective fantasies of both the state socialist world and the capitalist West (Buck-Morss 2000; see also Mackert and Geisthövel, this volume). Yet other approaches take a *longue durée* view to describe conceptualizations of productivity (see, for example, Burkhardt 1974). In the context of global history, they discuss the concept in combination with hypotheses such as that of the "Great Divergence" (Pomeranz 2000), in other words, the question of why Western European and other (in Pomeranz's study, Chinese) economic paths began to drift apart from 1750 on. Finally, the notion of productivity plays an important role in thought regarding the "industrious revolution", which focuses on the shifting orientation of family households with respect to consumption, work in the household, and wage labor (Hayami 1986; Vries 2008; Litvine 2014).

Genealogies of Productivity: A Body Historical Perspective

The present volume culls these avenues of research and then adds the perspective of body history to them. We are interested in how productivity became both a guiding concept of economic thought and the framing principle of a variety of economic practices. To that end, we systematically relate the history of this essentially contested concept to the body, starting from the view that the body is an interface, perhaps even *the* interface, connecting the various aspects and histories of productivity. Here the body is an anchor point; it is only in the body—moving, deploying its energy, expending itself—that notions of productivity take concrete form. Over the past 250 years, this function of the body as the linchpin of productivist assemblages of production has remained remarkably stable despite all the promises of rationalization and automation.

At the same time, the body has supplied a wealth of metaphors that have considerable traction for scholarly thinking on productivity and "economic life", to borrow a term coined by William Sewell (2010). Giving intelligible outlines to the complex relations of economic exchange becomes possible with a focus starting from the body. Abstract, macro contexts such as "the economy" become legible thanks to bodily metaphors of blood circulation, physical training, old-age frailty, and many more. As we will demonstrate in the following section that offers a brief history of the concept, in the eighteenth century, the changing figurations of the economic were carved out by bodily imagery, just as they are today. Put differently, the body is a puzzle picture made to capture another puzzle, the economy. Across very different notions of economic life, the body is both a microelement and a metaphor for macro contexts.

However, whereas the body is a reappearing linchpin in notions on productivity, it is not a continuous thread. Rather, the body lends itself to the project of testing genealogical perspectives on the economy. Genealogy tracks down history where no historicity was thought to exist—in the supposedly eternally valid, self-evident categories that structure our perception (Foucault [1971] 1977). Thus, genealogy does not seek origins, but rather uncovers the sequences and interconnections of discontinuous processes. In the past 250 years, the term "productivity" has not only had different meanings but no single guiding constant has linked those meanings. Unintended consequences acquired sudden importance, peripheral arenas gradually moved to the center, and problems once considered preeminent imperceptibly receded into the background. Consequently, the genealogical approach proceeds against the grain, so to speak, of the basic schemata of economic knowledge. This is why studies interested in the cultural and body histories of productivity have been able to focus on particular borderline cases of modern productivity or its antitheses. Work addressing such issues as fatigue (Rabinbach 1992), housework (Bock and Duden 1977), hunger (Vernon 2007), or the "vital minimum" (Simmons 2015) gives concrete shape to a surprising picture of modernity.[1]

The contributions in this volume also use case studies to explore historical situations that may seem to lie, upon initial examination, outside the remit of economic history: the integration of sick workers in the Weimar Republic, the body as a guarantor of quality and a source of danger in nineteenth-century French bakeries, colonial discourses on the "lazy natives" in West Africa, the politics of the calorie in the Progressive-Era US, population policy discourse in the seventeenth century, and so on. What brings these widely disparate case studies together is their focus on the body. Far from forming a homogeneous core, the body here appears as deeply embedded in historically specific assemblages right from the start—as a working, motivated, idle, tractable, reproducing body, as a body in need of protection. The physical form is always loaded with social expectations, always measured by means of particular knowledge practices.[2] Plus, individual bodies are always located in a complex interaction with particular arrangements of production.

Historians of labor have maintained a long-standing interest in the working body, and this volume hopes to build on their findings regarding the everyday life of workers (Lüdtke 1995; Canning 1996) and the production of "work" since the nineteenth century (Conrad et al. 2000; Kocka

2010; Bernet and Tanner 2015).[3] However, this is not a book about work and labor. Rather, our history of productivity approaches labor from the margins inward (see also Colin and Schößler 2013) and with a view to its epistemology—for productivity signifies specific processes of charging economic relations with meaning as they are observed, evaluated, and inserted into other interpretive grids. Practices oriented on productivity result in particular ways of formatting time, space, material resources, and measured effects. Thus, we regard labor as a relational force field within which the activity of bodies is harnessed to other societal discourses, institutions, and knowledge practices. In the past century and a half, for example, this includes the emergence of phenomena of "consumer and work society" (Bänziger 2015; see also Nolan 1994; Stanziani 2012), along with the scientization of production processes (Raphael 1996; Brückweh et al. 2012; Uhl 2014; Jarrige, this volume) and of life (Banta 1993; Tanner 1999), as well as the building of welfare-state institutions (Steinmetz 1993; Topalov 1994; Leimgruber 2008; Conrad 2013; Lengwiler 2015; Wadauer et al. 2015). Approached from this perspective, a history of productivity can bring to light the distinctions between "working" and "being productive", given that work may also possess a high value outside the productivist logic—*vide* work as an end to itself in the Benedictine *ora et labora*, or in the bourgeois work ethic of the nineteenth century (see, for example, Sarasin 2001; Ehmer and Lis 2009; Lis and Soly 2012; Verheyen 2014; Maß this volume).

Forms of knowledge production that surpass rationalization and labor science have been addressed by recent studies in the cultural history of economic life, which we also draw upon here. Such research problematizes the seemingly self-evident object that is "the economy",[4] and, in the process, "the economy" as a singular, reified concatenation of facts acquires history. According to some, that history is surprisingly short, beginning no earlier than the interwar period (Mitchell 2002).[5] Yet it is certainly not the prerogative of historians studying the twentieth century to defamiliarize "the economy" in its objective factuality. Studies of early modern economic thinking have located the formative figures of the eighteenth century, such as Adam Smith, Montesquieu, or the physiocrats, in the landscape of their own era's questions, irreversibly complicating the conventional founding narratives of economic science (see, for example, Hirschman 1977; Rothschild 2002; Schabas 2005; Cheney 2010).

Galvanized by historical epistemology and social studies of science, the cultural history of economic life has, of late, pursued a particular interest in the measuring techniques and modeling that frame and thereby produce the economic world (MacKenzie 2006; Preda 2009; Grimpe 2010; Morgan 2012; Stäheli 2013). The production of evidentiality by statistics, as exemplified by the gross domestic product, is especially important here (Tooze 2001; Desrosières 2002; Lepenies 2013; Speich Chassé 2013). However, ways of visualizing the economy, the establishment of measures and evaluation metrics—let us call them forms of metrological work—[6]are not carried out by economists or statisticians alone. They include the diverse and mostly inconspicuous "little tools of knowledge" employed by state agencies, workers, corporations, or consumers (Becker and Clark 2001).

The essays collected here link such metrological work to the body. They draw the dimensions of body history back into the current debate on a cultural history of economic life. For instance, when the "material transfers" (Groebner 2005, 82) by means of which bodies are traded on markets are analyzed historically, both body history and economic history gain in depth. Thus, investigating, for example, the relationship between slavery and capitalism shows what a massive agglomeration of physical violence, abstract market forces, and symbolic meaning-making was necessary for human bodies to be made into a commodity (Johnson 1999, 2013; Beckert 2014). Certainly, bodies functioned as commodities long before capitalism, and they continued to do so after chattel slavery's decline—e.g., in Europe, reaching from the mercenary system, to the early modern trade in human fat for pharmaceuticals manufacture, to the organ trade in the twenty-first century (Groebner 2005; Swanson 2014). Nevertheless, the history of capitalism retains a very particular connection to the commodification of bodies and of their activity.[7] This is because the metrological work of commodification in capitalism has a peculiar, twofold relationship to the body. On the one hand, in capitalism, the economy has been described as a magnitude on its own account, devoid of corporeality: thus, by the moment when the economy became envisioned as a separate sphere—a moment in the European eighteenth century, when the economy was already capitalist—it was described as an opaque object (Buck-Morss 1995). For instance, to recapitulate Karl Marx's famous formula, the movement from M to M' does not involve a visible dimension. And yet, on the other hand, it was because of this very opacity that, time and again, the economy took on bodily figurations (Vogl 2014). A conceptual history can demonstrate quite vividly how this relationship to the body shaped the changing topography of "productivity".

Conceptual History: The Corporeal Foundations of Modern Productivity Discourses[8]

Before early modern scholars addressed the semantic field of "productivity", the verb "to produce" was commonly used in jurisprudence, meaning to submit or exhibit testimonial documents in court. At the same time, premodern merchants used the noun "produce" in a simple arithmetic sense to designate the calculated result of a commercial operation (Hentschel 1984, 2–5). In the course of the eighteenth century, the term was increasingly employed in a more systematic way, but before its economic meaning became hegemonic in the nineteenth century, it coexisted with aesthetic and philosophical conceptualizations. For Immanuel Kant, Wolfgang von Goethe, and other protagonists of Enlightenment and Romanticism, for instance, the concept of artistic productivity was closely linked to the creativity of the "genius". Kant developed a notion of "productive" and "reproductive" imagination (König 1989, 1421–3); as a bourgeois observer of himself, Goethe reflected on the conditions of his "productivity" as a writer (1423, 1433). Although the extent to which these concepts of artistic creativity were influenced by contemporary economic thinking remains unclear, it seems obvious that they resemble some of the definitions of productivity developed in economics during the nineteenth century and especially in the second half of the twentieth century (Reckwitz 2012; Blaschke 2013; Haunschild 2013; Knoblauch et al. 2014; and see the following section). Not least, productivity has become one of the central concepts of post-structuralist historiography and the cultural sciences in general, which, to date, have hardly been historicized themselves.[9]

In the field of political economy, the term "productivity" first gained prominence in the writings of the French physiocrats.[10] François Quesnay's cyclical model of economy, as set out in his *Tableau économique* of 1758, for instance, was rooted both in his medical knowledge of the human body, particularly of blood circulation, and in early modern agricultural knowledge that was profoundly influenced by the problem of "subsistence". Hence, productivity was a quality of material, natural processes that were described in terms of "fertility" and "generation" (Christenson 1994, 271). In the strict sense, only the soil was "productive". To a much lesser extent, the laboring classes, occupied with growing and harvesting these spontaneous gifts of "mother nature", were considered to be "productive" as well. Similarly, Quesnay (1758) distinguished between "productive" and "sterile classes", and included manufacturing labor in the latter category. In so doing, he indirectly contrasted "sterile" with "fertile". Thus, fertility turned out to be the defining criterion of "being productive".[11]

Adam Smith further developed and enriched the economic concept of productivity in his *Inquiry into the Nature and Causes of the Wealth of Nations*, published in 1776. In direct opposition to the physiocrats, this monumental reference point in the controversial discussions among late eighteenth- and nineteenth-century European economists places labor at the core of its definition of productivity.[12] Smith's definition of "productive" as opposed to "unproductive" labor displaces and transforms the physiocrats' distinction. For him, there are two distinctive features of productive labor: first, labor is productive to the extent that it "adds to the value of the object upon which it is bestowed" and second, it is productive when it "fixes and realizes itself in some particular subject or vendible commodity" (Smith 1776, 270). With this definition of productive labor, there is a striking lack of any reference to the body or bodily strains, which had been constitutive for physiocratic thinking. Throughout his book, however, Smith constantly uses the metaphor of "hands", which implicitly adds a bodily dimension to his economic theory. First, "hands" serve as a metaphor for control and power of disposal. Second, and even more intriguingly, the term consistently signifies "laborers", "industrious people", or "workmen", thus reducing the laboring population to their physical dimension, as a part of the body. Furthermore, Smith subsumes "the labor of some of the most respectable orders in the society" under the category of "unproductive labor". Hence, the sovereign, public and domestic servants, and the army and navy found themselves equated with "churchmen, lawyers, physicians, men of letters of all kinds; players, buffoons, musicians, opera-singers, opera-dancers, etc." All these professions, Smith argues, commonly share that their work neither creates value nor realizes itself in vendible objects. Rather, it "perishes in the very instant of its production" (271).

It is powerful evidence of the centrality accorded to *The Wealth of Nations* in economic discourse by the middle of the nineteenth century that some of these allegedly unproductive laborers tried to prove their usefulness (see also Burkhardt 1974, 280–3). Addressing Smith's verdict on public servants' unproductive labor in general, and written in the context of the German national unification movement, Friedrich List's 1841 book *The National System of Political Economy* developed a theory of "productive forces" that explicitly opposed the "theory of value" outlined in *The Wealth of Nations*. His theory allowed List to transform the distinction of productive versus unproductive labor yet again and to stress the economic necessity of government: "Those who raise pigs, and those who manufacture bag-pipes or pills, are indeed productive", he ironically alluded to Smith, "but the instructors of youth and of manhood, musicians, virtuosos, physicians, judges, and statesmen, are productive in a much higher degree. The former produce exchangeable values; the latter, productive power" (List [1841] 1856, 221). Like the similar arguments of Adam Heinrich Müller (1809), List's theory was one of economic government in general as well as a plea for national economic policies in particular. It enlarged the economic concept of productivity inasmuch as, alongside the natural fertility of the soil and the materialization of labor in exchangeable and vendible objects, it added the production and maintenance of economic productivity by the state as a third dimension.

A critique of Smith's concept of productivity and its reformulations by British, French, and German scholars formed the starting point for Karl Marx's own theorization of productivity. He rejected Smith's distinction of productive/unproductive labor, and the classification of social roles and professions that it entailed, as bourgeois ideologies (Marx [1862–1863] N.D., 35–7). From his perspective, any labor was productive, provided that it was paid and contributed to the creation of surplus value and the reproduction of capital. The accumulation and reproduction of capital was dependent on the degree of productivity of societal labor, as well as on the "intensity" of labor. In Marx's vision of capitalist production relations as detailed in *Das Kapital* (1867), "productivity" was a quality of wage labor only and, even more importantly, a significant feature of societal organization and classification under capitalist command (Marx [1867] 1976).

However, as some have argued, Marx's conceptualization of labor shifted over the years (see Althusser and Balibar [1965] 1970, part II). In his early writings, labor was a transhistorical metabolism between humans and nature. In contrast, the notion of "labor power", arguably inspired by thermodynamics and first introduced in the *Grundrisse* of 1857, enabled a new conceptualization (Burkett and Foster 2006). Living labor—that is, a worker's body and living time—was now opposed to abstract labor by the intermediation of labor power, an energetic total sum that was traded as a commodity on the labor market, enacted in the labor process, and, thus, enabled the creation of surplus value. Consequently, labor was reconfigured as an essentially historical category, and as just one specific organization in a vast array of possible arrangements between human and nonhuman forces (Marx [1857] 1973). The hybrid monsters and specters, the suction by which dead labor feeds off living labor, not only document the importance of organicist metaphors in Marx's writing but also underline the novelty of the figures and situations that Marx thought were arising in the process of accelerated productivity in industrial capitalism (Tomba 2013).

With the emergence of the hygienic and social reform movements and the formation of labor science around 1870, the body of the industrial laborer was drawn into economic discourse in a different way, although no less powerfully. As Anson Rabinbach (1992; see also Osietzki 1998) has pointed out, the afore-mentioned thermodynamic model of labor developed by Hermann von Helmholtz, and its subsequent use for the exploration and measurement of the working body, enabled an objectivist understanding of labor and the productivity of the human body. By the beginning of the twentieth century, the worker's body was surrounded by an entire technical apparatus designed to measure and calculate its productivity. That productivity was conceived in terms of effort and exhaustion; new methods, such as ergography, served to measure work intensity and allowed different phases of strain and fatigue to be identified. Instruments, such as the calorimeter, were invented to measure and calculate diets for the industrial worker that optimized the relation between calorific input and energetic output of the "human motor" (see Mackert, this volume). At the end of World War I, some labor scientists even started to analyze the productivity of "disabled" bodies (see Amar 1916; 1918; Geisthövel, this volume).

At the same time, several leading German political economists dismissed the concept of productivity as largely useless precisely because it had been subject to such a variety of theoretical approaches, and, thus, seemed to have lost its analytical value.[13] At the 1909 annual meeting of the leading German social reform association, the Verein für Socialpolitik (Association for Social Policy), in Vienna, Max Weber ([1909] 1988) vehemently criticized a contribution on whether, and to what extent, productivity should be treated as the main purpose of economic systems. Against the background of intense debates on "municipal socialism" and productivist visions of political economy, Weber postulated that the question itself was fallacious.

Instead of mixing ethics with science and thus infringing on science's neutrality, he argued that the concept could be useful for economics and empirical social research only as a mere technical term.[14]

Weber's insistence on the technical character of the concept, and his propositions regarding its measurement and calculation, pointed in a direction that French, American, and British economists were already pursuing (Knortz 2010). The term "productivity" occasionally appears in the US debate, especially in Frederick Winslow Taylor's (1911) influential *The Principles of Scientific Management*, although far less often than "efficiency". Taylor himself argued that the notion of efficiency was borrowed from the engineering sciences and fostered a mechanical understanding of productivity as an abstract relationship between input and output (see Efficiency 1934, 437). Taylor's essay was widely read in German, but with some differences: as a comparison of the original text to its German translation has recently shown, both terms—productivity and efficiency—entailed physiological connotations in German that had not been present in the English version (Knortz 2010, 36). In France, economist Albert Aftalion (1911) published a long article developing a concept of productivity that more closely resembled the principles of scientific management depicted by Taylor. He defined productivity as a relationship between the input and "physical output in certain objects in a given time span". Much of Aftalion's article strove to define the formal variables and the right factors (labor, capital) and scales (firms, branches, national economy) that should be included in the calculation.[15]

During the first half of the twentieth century, especially in the context of debates on reconstruction policies following the two world wars, this technical notion of economic productivity clearly became hegemonic.[16] In the writings of the influential French engineer and economist Jean Fourastié (1959) and his advisory work for the OECD, productivity was a crucial feature for evaluating the scientific and technological progress of any national economy or its leading economic sectors (see Boulat 2008). However, during the *trente glorieuses*, the concept itself was not really at stake, even if some protagonists, such as the German economist Gerhart E. Reuss (1960, 27), working on behalf of the West German government, still lamented the vagueness that prevailed. Central to the discussion, instead, were the modes and models of statistical measurement and the macroeconomic calculation of "productivity indices" with respect to economic policies and international comparison (Speich Chassé 2013; see also Schmelzer 2016 and Held, this volume).

Starting in the 1960s, in the context of debates on deindustrialization and the expansion of the service sector, microeconomic "human capital" theories reintegrated the human body into conceptualizations on productivity, although in different ways (see Bernet and Gugerli 2011). For instance, Gary Becker (1962) and Theodore W. Schultz (1962) shifted the focus from production processes on the level of the firm to the individual laborer's consumption and leisure preferences. Just like the labor scientists of the late nineteenth century, these authors depicted productivity as a quality of the individual worker, but they introduced two major differences: on the one hand, the task of increasing productivity was imposed on the individual laborer so that marginal productivity depended on investments in education and individual self-improvement; on the other hand, the very possibility of raising productivity and self-improvement was limited by the worker's natural potentials, for example, his or her inherited bodily and genetic constitution (see Foucault [1978] 2004, 220–44). However, emerging motivation psychology increasingly discarded these restraints. As Lukas Held (this volume) argues, fatigue was soon no longer conceived of as an absolute threshold of productive labor—it became a transitional stage that could be subject to individual training (see also Donauer 2013).

In light of this short sketch of the modern debates on productivity, we can draw three conclusions, which also form the general argument of the present volume. First, the scope of the essentially contested concept of productivity was progressively enlarged from the middle of the eighteenth and especially in the course of the nineteenth century, to the extent that it was able to become a notion crucial to the understanding of any economic activity in the twentieth century. Second, the economic sciences have clearly loosened productivity's connections to its material foundations (soil, bodies, objects) in the process of theorization and modeling for statistical purposes. Lastly, however, the material and corporeal dimension of productivity, nonetheless, did not vanish completely. On the contrary, its influence and impact is still palpable today, especially in social economics and its empirical field studies.

The Analytical Frame: "Regimes of Productivity"

This tentative survey indicates that both the material dimensions of productivity—the soil, vendible commodities, and so on—and the boundaries between productive and unproductive labor have been subject to constant transformation, and controversies on productivity have always also been debates about societal inclusions and exclusions. They carry valuations and devaluations of certain bodies, objects and agents, and professions and classes. The semantics of productivity thus turns out to be a rationale of modern biopolitics (Foucault [1978] 2004). Consequently, it is useful to go a step further and approach the histories of productivity from a more capacious perspective than the focus on historical semantics and influential thinkers, while still remaining aware of their performative impacts.

This is why this volume proposes to analyze "regimes of productivity" rather than concepts of productivity alone. The notion of regime, influenced by the work of Michel Foucault, is both constructivist and genealogical. It points to the changing conditions of possibility as well as the ongoing interplay of discourses and practices that constitute basic elements of economic life. Instead of postulating a single concept of productivity or generally binding historical stages, an analysis of regimes of productivity invites us to attend to the border traffic between fields of knowledge and domains of practice. "Energy" and "performance" are well-studied examples, having oscillated since the nineteenth century between physics, economics, physiology, and ergonomics (Rabinbach 1992; Vatin 1993). In industrial praxis, they helped to establish specific dispositions of spaces, machines, and bodies, while simultaneously producing bodies and informing selves (Braun 1991; Sarasin 2001). An approach of this kind, then, investigates mid-scale phenomena as a point of departure for contributing to the history of the modern economy. By systematizing the analytical approaches of the chapters that follow, one can discern at least four points from which such a project might proceed: bodies, things/objects, knowledge/concepts, and translation/assembly. Obviously, these heuristic starting points reflect recent debates in cultural history.

To start with the angle of body history, as the previous section indicates, the involvement of bodies has been central to regimes of productivity since they emerged in the European societies of the seventeenth and eighteenth centuries. Because bodies are historically contingent artifacts, it is useful to specify the bodies at stake in concrete settings—working bodies, bodies giving birth, trained body parts, and so forth—and to consider a range of qualities and processes of discursive qualification and ascription. For instance, the pivotal distinction productive/unproductive has always been concretized. To unpack the trajectories leading to such binary oppositions, students of body history make use of process-oriented concepts, such as gendering, individuation/totalization, racialization, and more. Oftentimes, such qualifications are entangled in manifold ways, as the chapters in this volume indicate.[17]

It is equally important to look in detail at the objects and processes of objectification within particular regimes of productivity.[18] We focus on the concepts of productivity at stake within such processes. In this way, the discussion on whether only human bodies can become productive bodies is restaged as an essentially historical one. Obviously, it is almost impossible to separate working subjects from their tools and means of production, from the simple hammer to the "cube farm". However, analyzing specific regimes of productivity means asking after the concrete operations through which these objects were deemed to be productive or not.

Another dimension of the emergence and transformation of regimes of productivity is knowledge, in a variety of forms. An important example is the disciplined knowledge of the emerging sciences in the nineteenth and twentieth centuries, especially physiology, a science that proved crucial to the conceptualization of productivity in the age of industrialization. Yet smaller, locally limited agencies, or subversive practices, can also elaborate and translate knowledge. Furthermore, discussion of the body and its productivity has often implied knowledge of the self, especially in the twentieth century.

Finally, the arrangements of bodies in built environments give rise to contexts that some scholars have referred to as "technopolitics" (Hecht 1998;

Mitchell 2002; Simmons 2015). Reflection on technopolitics has a long history, ranging from Marx's deliberations in his *Fragment on Machines* ([1857] 1973, 690–714) to Donna Haraway's *A Manifesto for Cyborgs* (1985). That history has been enriched in recent years by key terms such as "network" or "association". The focus is on the dynamics of translation, connection, interaction, friction, and resistance between bodies, materialities, media, and symbols.[19] As the reference to "politics" indicates, what is at stake here is not some mechanical determinism, but contested processes that play out within social and epistemic registers. We are concerned with avoiding setting any conceptual a priori—whether those of crude materiality or of technological determinism, or conversely, indeed of "social causes", ultimately understood tautologically—but, instead, endeavor to keep our research questions open. "Politics" also point to the contradictory unity of different, close and distant effects. It is here, in our opinion, that the analytical power of the concept of "regime" is to be found, embedding politics into distinct and describable assemblages.

Against this backdrop, the notion of regime allows us to analyze the variety of technologies of power—of government and self-conduct—that are at work in the constitution and materialization of historically contingent and heterogeneous assemblages and within which distinct regimes of productivity have been prevalent. Using and testing the analytical frame of regimes, the chapters in this volume approach the histories of productivity sideways, so to speak. They explore the borderlands of work and nonwork, the metrics and the performative impacts of measurement, and the translations of an economic rationale, asking how bodies were made and how they came to be seen as productive.

Notes

1 This account owes much to Dana Simmons (2015, 7).

2 We draw the term "knowledge practices" from Marilyn Strathern's (1988) comparative anthropology of gender relations, a relational ontology that has not yet attracted substantial scholarly attention from historians.

3 An overview of the prolific "new" labor history is given by Neville Kirk (2012), Jörg Neuheiser (2013), Lex Heerma van Voss (2013), and Kim Christian Priemel (2014), among others.

4 "Problematization" as a critical form of analysis looks at the practical horizons within which a problem can become intelligible. For an introduction to the notion of problematics as set out by Gaston Bachelard, see Patrice Maniglier (2012). For recent contributions to a history of economic knowledge, see Monika Dommann et al. (2014).

5 Different views on periodization are proposed by Manu Goswami (2004, 215–24), for example; see also Tomothy Mitchell (2011, 125–7) and Christof Dejung et al. (2014).

6 We borrow the term from Michael Wintroub (2015, 1186); furthermore, on the significance of metrology for a global history of science, see Simon Schaffer (2015).

7 As new work has shown, the spectrum of figures and subjects concerned is by no means limited to slaves alone, but includes a multitude of historical actors, for instance, clerks as well (see Zakim 2012).

8 This overview owes much to the seminal article by Johannes Burkhardt (1974).
9 Some remarks by Philipp Felsch (2015, 127–30) may serve as a starting point.
10 However, it has recently been argued that although the notion itself did not yet exist, some of its elements can already be found in mercantilistic thought (Laude and Heß 2008b, 19; Buchner 2008; 2009). For the early modern context, see also Justus Nipperdey, this volume.
11 For premodern and early modern economic thought, see also Corinna Laude and Gilbert Heß (2008a) and Justus Nipperdey (this volume).
12 Contemporary English authors used the term "productiveness" in quite a similar way; see, for example, Josiah Conder (1833).
13 Werner Sombart (1928) tackled the problem starting from its semantic aspects and identified three different concepts of productivity, subdivided into several distinct notions. In the end, he listed no fewer than ten different meanings of the concept in the economic discourse of his day. On productivity in contemporary German sociology, see also Andreas Langenohl (2013).
14 Max Weber ([1908] 1995, 140) dedicated himself, nonetheless, to the analysis of the "psycho-physics of labor", and made various attempts to survey labor productivity on the ground. His report does not use the term productivity—except for a single sentence, in which he picks up the term "productivity of the brain" and alludes to Abbé, a contemporary French expert in labor science.
15 Régis Boulat (2002, 2006, 2008) has studied the French debate on productivity in the early to mid-twentieth century. See also the seminal contribution by Patrick Fridenson (1987).
16 For the German-speaking debate in the 1920s, see Werner Sombart (1928) and Frieda Wunderlich (1926).
17 In particular, see the chapters by Cassandra Mark-Thiesen, Tristan Oestermann, Nina Mackert, Alexa Geisthövel, and Monika Streule.
18 Although he does not explicitly mention productivity, see Frank Trentmann (2009) for some seminal thoughts in this respect.
19 On "friction", a term at least as important as the much-discussed "circulation", see Anna Lowenhaupt Tsing (2005).

Bibliography

Aftalion, Albert. 1911. "Les trois notions de la productivité et les revenus." *Revue d'économie politique* 25 (1): 145–84; 345–69.

Althusser, Louis, and Etienne Balibar. (1965) 1970. *Reading Capital*. Translated by Ben Brewster. London: New Left Books.

Amar, Jules. 1916. *La prothèse et le travail des mutilés*. Paris: Dunod & Pinat.

———. 1918. *The Physiology of Industrial Organisation and the Re-employment of the Disabled*. Translated by Bernard Miall. London: Library Press.

Banta, Martha. 1993. *Taylored Lives: Narrative Productions in the Age of Taylor, Veblen, and Ford*. Chicago: The University of Chicago Press.

Bänziger, Peter-Paul. 2015. "Von der Arbeits- zur Konsumgesellschaft? Kritik eines Leitmotivs der deutschsprachigen Zeitgeschichtsschreibung." *Zeithistorische Forschungen* 12 (1): 11–38.

Becker, Gary S. 1962. "Investment in Human Capital: A Theoretical Analysis." *The Journal of Political Economy* 70 (5): 9–49.

Becker, Peter, and William Clark, eds. 2001. *Little Tools of Knowledge: Historical Essays on Academic and Bureaucratic Practices*. Ann Arbor: University of Michigan Press.

Beckert, Sven. 2014. *The Empire of Cotton: A Global History*. New York: A. Knopf.

Bernet, Brigitta, and David Gugerli. 2011. "Sputniks Resonanzen: Der Aufstieg der Humankapitaltheorie im Kalten Krieg—eine Argumentationsskizze." *Historische Anthropologie* 19 (3): 433–46.

Bernet, Brigitta, and Jakob Tanner, eds. 2015. *Ausser Betrieb: Metamorphosen der Arbeit in der Schweiz*. Zurich: Limmat.

Blaschke, Bernd. 2013. "Das nennen Sie Arbeit?! Produktivitätsdiskurse der Bohème vom 19. bis zum 21. Jahrhundert." In *Das nennen Sie Arbeit? Der Produktivitätsdiskurs und seine Ausschlüsse*, edited by Nicole Colin and Franziska Schößler, 245–66. Heidelberg: Synchron.

Bock, Gisela, and Barbara Duden. 1977. "Arbeit aus Liebe—Liebe als Arbeit: Zur Entstehung der Hausarbeit im Kapitalismus." In *Frauen und Wissenschaft: Beiträge zur Berliner Sommeruniversität für Frauen, Juli 1976*, edited by Gruppe Berliner Dozentinnen, 118–99. Berlin (FRG): Courage.

Boulat, Régis. 2002. "Le concept de productivité en France de la Première Guerre mondiale aux années soixante." *Travail et Emploi* 91: 43–56.

———. 2006. "La productivité et sa mesure en France (1944–1955)." *Histoire & Mesure* 21 (1): 79–110.

———. 2008. *Jean Fourastié, un expert en productivité: La modernisation de la France (années trente—années cinquante)*. Besançon: Presses universitaires de Franche-Comté.

Braun, Rudolf. 1991. "The 'Docile' Body as an Economic-Industrial Growth Factor." In *Favorites of Fortune: Technology, Growth, and Economic Development Since the Industrial Revolution*, edited by Patrice Higonnet, David S. Landes, and Henry Rosovsky, 120–41. Cambridge, MA: Harvard University Press.

Brückweh, Kerstin, Dirk Schumann, Richard F. Wetzell, and Benjamin Ziemann, eds. 2012. *Engineering Society: The Role of the Human and Social Sciences in Modern Societies, 1880–1980*. Basingstoke: Palgrave Macmillan.

Buchner, Thomas. 2008. "Arbeit, Ordnung—Produktivität? Ein Vergleich von niederländischem Merkantilismus und deutschsprachigem Kameralismus im 17. Jahrhundert." In *Konzepte von Produktivität im Wandel vom Mittelalter in die Frühe Neuzeit*, edited by Corinna Laude and Gilbert Heß, 315–45. Berlin: Akademie Verlag.

———. 2009. "Perceptions of Work in Early Modern Economic Thought: Dutch Mercantilism and Central European Cameralism in Comparative Perspective." In *The Idea of Work in Europe from Antiquity to Modern Times*, edited by Josef Ehmer and Catharina Lis, 191–213. Farnham: Ashgate.

Buck-Morss, Susan. 1995. "Envisioning Capital: Political Economy on Display." *Critical Inquiry* 21: 434–67.

———. 2000. *Dreamworld and Catastrophe: The Passing of Mass Utopia in East and West*. Cambridge, MA: MIT Press.

Burkett, Paul, and John Bellamy Foster. 2006. "Metabolism, Energy and Entropy in Marx's Critique of Political Economy: Beyond the Podolinsky Myth." *Theory and Society* 35: 109–56.

Burkhardt, Johannes. 1974. "Das Verhaltensleitbild 'Produktivität' und seine historisch-anthropologische Voraussetzung." *Saeculum* 25: 277–89.

Canning, Kathleen. 1996. *Languages of Labor and Gender: Female Factory Work in Germany, 1850–1914*. Ithaca: Cornell University Press.

Cheney, Paul. 2010. *Revolutionary Commerce: Globalization and the French Monarchy*. Cambridge, MA: Harvard University Press.

Christenson, Paul P. 1994. "Fire, Motion, and Productivity: The Proto-Energetics of Nature and Economy in Francois Quesnay." In *Natural Images in Economic Thought: "Markets Read in Tooth and Claw"*, edited by Philip Mirowski, 249–88. Cambridge, MA: Cambridge University Press.

Colin, Nicole, and Franziska Schößler, eds. 2013. *Das nennen Sie Arbeit? Der Produktivitätsdiskurs und seine Ausschlüsse*. Heidelberg: Synchron.

Conder, Josiah. 1833. *Wages or the Whip: An Essay on the Comparative Cost and Productiveness of Free and Slave Labour*. London: Hatchard and Son.

Conrad, Christoph. 2013. "Was macht eigentlich der Wohlfahrtsstaat? Internationale Perspektiven auf das 20. und 21. Jahrhundert." *Geschichte und Gesellschaft* 39: 555–92.

Conrad, Sebastian, Elisio Macamo, and Bénédicte Zimmermann. 2000. "Die Kodifizierung der Arbeit: Individuum, Gesellschaft, Nation." In *Geschichte und Zukunft der Arbeit*, edited by Jürgen Kocka and Claus Offe, 449–75. Frankfurt am Main: Campus.

Dejung, Christof, Monika Dommann, and Daniel Speich Chassé, eds. 2014. *Auf der Suche nach der Ökonomie: Historische Annäherungen*. Tübingen: Mohr Siebeck.

Desrosières, Alain. (1993) 2002. *The Politics of Large Numbers: A History of Statistical Reasoning*. Translated by Camille Naish. Cambridge, MA: Harvard University Press.

Dommann, Monika, Daniel Speich Chassé, and Mischa Suter. 2014. "Einleitung: Wissensgeschichte ökonomischer Praktiken." *Berichte zur Wissenschaftsgeschichte* 37 (2): 107–11.

Donauer, Sabine. 2013. "Emotions at Work—Working on Emotions: The Production of Economic Selves in Twentieth-Century Germany." PhD diss., Free University, Berlin.

"Efficiency". 1934. In *Encyclopaedia of the Social Sciences*. Vol. 5, edited by Edwin R.A. Seligman, 437–39. New York: MacMillan.

Ehmer, Josef, and Catharina Lis, eds. 2009. *The Idea of Work in Europe from Antiquity to Modern Times*. Farnham: Ashgate.

Felsch, Philipp. 2015. *Der lange Sommer der Theorie: Geschichte einer Revolte 1960–1990*. München: C.H. Beck.

Field, Alexander J. 2008. "Productivity." In *The Concise Encyclopedia of Economics*, edited by David R. Henderson. Indianapolis. Accessed September 20, 2014. www.econlib.org/library/Enc/Productivity.html.

Foucault, Michel. (1971) 1977. "Nietzsche, Genealogy, History." In *Language, Counter-Memory, Practice*, edited by Donald Bouchard, translated by D. F. Bouchard and Sherry Simon, 139–64. Ithaca, NY: Cornell University Press.

———. (1978) 2004. *Naissance de la biopolitique*. Paris: Gallimard/Seuil.

Fourastié, Jean. 1959. *La productivité*. Paris: PUF.

Fridenson, Patrick. 1987. "Un tournant taylorien de la société française (1904–1918)." *Annales. Histoire, Sciences Sociales* 42 (5): 1031–60.

Gallie, Walter B. 1956. "Essentially Contested Concepts." *Proceedings of the Aristotelian Society* 56: 167–98.

Goswami, Manu. 2004. *Producing India: From Political Economy to National Space*. Chicago: The University of Chicago Press.

Grimpe, Barbara. 2010. *Ökonomie sichtbar machen: Die Welt nationaler Schulden in Bildschirmgröße: Eine Ethnographie*. Bielefeld: Transcript.

Groebner, Valentin. 2005. "Körper auf dem Markt: Söldner, Organhandel und die Geschichte der Körpergeschichte." *Mittelweg 36: Zeitschrift des Hamburger Instituts für Sozialforschung* 14: 69–84.

Haraway, Donna. 1985. "A Manifesto for Cyborgs: Science, Technology, and Socialist Feminism in the 1980s." *Socialist Review* 15 (2): 65–107.

Haunschild, Axel. 2013. "Emotional und Aesthetic Labor: Zur gesellschaftlichen Konstruktion der Produktivität von Dienstleistungsarbeit." In *Das nennen Sie Arbeit? Der Produktivitätsdiskurs und seine Ausschlüsse*, edited by Nicole Colin and Franziska Schößler, 297–312. Heidelberg: Synchron.

Hayami, Akira. 1986. "A Great Transformation: Social and Economic Change in Sixteenth and Seventeenth Century Japan." *Bonner Zeitschrift für Japanologie* 8: 3–13.

Hecht, Gabrielle. 1998. *The Radiance of France: Nuclear Power and National Identity After World War II*. Cambridge, MA: MIT Press.

Hentschel, Volker. 1984. "Produktion, Produktivität." In *Geschichtliche Grundbegriffe*. Vol. 5, edited by Otto Brunner, Werner Conze, and Reinhart Koselleck, 1–26. Stuttgart: Klett-Cotta.

Hirschman, Albert. 1977. *The Passions and the Interest: Political Arguments for Capitalism Before Its Triumph*. Princeton, NJ: Princeton University Press.

Johnson, Walter. 1999. *Soul by Soul: Life Inside the Antebellum Slave Market*. Cambridge, MA: Harvard University Press.

———. 2013. *River of Dark Dreams: Slavery and Empire in the Cotton Kingdom*. Cambridge, MA: Harvard University Press.

Kirk, Neville. 2012. "Taking Stock: Labor History During the Past Fifty Years." *International Labor and Working-Class History* 82: 156–73.

Knoblauch, Hubert, Mark Jacobs, and René Tuma, eds. 2014. *Culture, Communication, and Creativity: Reframing the Relations of Media, Knowledge, and Innovation in Society*. New York: Peter Lang.

Knortz, Heike. 2010. "Die Entwicklung des modernen Begriffs von der Arbeitsproduktivität: Ein Beitrag zur Geschichte der deutschen Betriebswirtschaftslehre." *Zeitschrift für Unternehmensgeschichte* 55 (1): 31–51.

Kocka, Jürgen, ed. 2010. *Work in a Modern Society: The German Historical Experience in Comparative Perspective*. New York: Berghahn.

König, René. 1989. "Produktion, Produktivität." In *Historisches Wörterbuch der Philosophie*. Vol. 7, edited by Joachim Ritter, Karlfried Gründer, and Gottfried Gabriel, 1418–37. Basel: Schwabe.

Langenohl, Andreas. 2013. "Finanzmarktdiskurs und Produktivitätsdiskurs in Deutschland um 1900: Anmerkungen zu Werner Sombart, Max Weber und Georg Simmel." In *Das nennen Sie Arbeit? Der Produktivitätsdiskurs und seine Ausschlüsse*, edited by Nicole Colin and Franziska Schößler, 87–102. Heidelberg: Synchron.

Laude, Corinna, and Gilbert Heß, eds. 2008a. *Konzepte von Produktivität im Wandel vom Mittelalter in die Frühe Neuzeit*. Berlin: Akademie Verlag.

———. 2008b. "Einleitung." In *Konzepte von Produktivität im Wandel vom Mittelalter in die Frühe Neuzeit*, edited by Corinna Laude and Gilbert Heß, 7–25. Berlin: Akademie.

Leimgruber, Matthieu. 2008. *Solidarity Without the State? Business and the Shaping of the Swiss Welfare State, 1890–2000*. Cambridge: Cambridge University Press.

Lengwiler, Martin. 2015. "Cultural Meanings of Social Security in Postwar Europe." *Social Science History* 39 (1): 85–106.

Lepenies, Philipp. 2013. *Die Macht der einen Zahl: Eine politische Geschichte des Bruttoinlandsprodukts*. Berlin: Suhrkamp.

Lis, Catharina, and Hugo Soly. 2012. *Worthy Efforts: Attitudes to Work and Workers in Pre-Industrial Europe*. Leiden and Boston: Brill.

List, Friedrich. (1841) 1856. *The National System of Political Economy*. Translated by G. A. Matile. Philadelphia: Lippincott.

Litvine, Alexis D. 2014. "The Industrious Revolution, the Industriousness Discourse, and the Development of Modern Economies." *The Historical Journal* 57 (2): 531–70.

Lüdtke, Alf, ed. 1995. *The History of Everyday Life: Reconstructing Experiences and Ways of Life*. Princeton, NJ: Princeton University Press.

MacKenzie, Donald. 2006. *An Engine, Not a Camera: How Financial Models Shape Markets*. Cambridge, MA: MIT Press.

Maier, Charles. 1977. "The Politics of Productivity: Foundations of American International Economic Policy After World War II." *International Organization* 31 (4): 607–33.

Maniglier, Patrice. 2012. "What Is a Problematic?" *Radical Philosophy* 173: 21–3.

Marx, Karl. (1857) 1973. *Grundrisse: Foundations of the Critique of Political Economy*. Translated by Martin Nicolaus. London: Penguin.

———. (1862–1863). N.D. *Theories of Surplus Value*. Vol. 1, translated by Emile Burns. Moscow: Foreign Languages Publishing House.

———. (1867) 1976. *Capital: A Critique of Political Economy*. Vol. 1, translated by Ben Fowkes. London: Penguin.

Meadows, Donella H., Dennis L. Meadows, Jørgen Randers, and William W. Behrens III. 1972. *The Limits to Growth: A Report for the Club of Rome's Project on the Predicament of Mankind*. New York: Universe Books.

Mitchell, Timothy. 2002. *Rule of Experts: Egypt, Techno-Politics, Modernity*. Berkeley: University of California Press.

———. 2011. *Carbon Democracy: Political Power in the Age of Oil*. London: Verso.

Morgan, Mary S. 2012. *The World in the Model: How Economists Work and Think*. Cambridge: Cambridge University Press.

Müller, Adam Heinrich. 1809. *Die Elemente der Staatskunst*. Volumes 1–3. Berlin: Sander.

Neuheiser, Jörg. 2013. "Arbeit zwischen Entgrenzung und Konsum: Die Geschichte der Arbeit im 20: Jahrhundert als Gegenstand aktueller zeithistorischer und sozialwissenschaftlicher Studien." *Neue Politische Literatur* 58 (3): 421–48.

Nolan, Mary. 1994. *Visions of Modernity: American Business and the Modernization of Germany*. New York: Oxford University Press.

OECD, Federal Statistical Office, eds. 2008. *Productivity Measurement and Analysis*. Paris: OECD Publishing.

Osietzki, Maria. 1998. "Körpermaschinen und Dampfmaschinen: Vom Wandel der Physiologie und des Körpers unter dem Einfluß von Industrialisierung und Thermodynamik." In *Physiologie und industrielle Gesellschaft: Studien zur Verwissenschaftlichung des Körpers im 19. und 20. Jahrhundert*, edited by Philipp Sarasin and Jakob Tanner, 313–46. Frankfurt am Main: Suhrkamp.

Pomeranz, Kenneth. 2000. *The Great Divergence: China, Europe, and the Making of the Modern World Economy*. Princeton: Princeton University Press.

Preda, Alex. 2009. *Framing Finance: The Boundaries of Markets and Modern Capitalism*. Chicago: The Chicago University Press.

Priemel, Kim Christian. 2014. "Heaps of Work: The Ways of Labor History." *H-Soz-u-Kult* 23 January 2014. Accessed February 24, 2016. http://www.hsozkult.de/literaturereview/id/forschungsberichte-1223.

Quesnay, François. 1758. *Tableau économique, et maximes générales du gouvernement économiques*. Versailles.

Rabinbach, Anson. 1992. *The Human Motor: Energy, Fatigue, and the Origins of Modernity*. Berkeley: University of California Press.

Raphael, Lutz. 1996. "Die Verwissenschaftlichung des Sozialen als methodische und konzeptionelle Herausforderung für eine Sozialgeschichte des 20. Jahrhunderts." *Geschichte und Gesellschaft* 22: 165–93.

Reckwitz, Andreas. 2012. *Die Erfindung der Kreativität: Zum Prozess gesellschaftlicher Ästhetisierung*. Frankfurt am Main: Suhrkamp.

Reuss, Gerhart E. 1960. *Produktivitätsanalyse: Ökonomische Grundlagen und statistische Methodik*. Basel: Kyklos.

Rothschild, Emma. 2002. *Economic Sentiments: Adam Smith, Condorcet, and the Enlightenment*. Cambridge, MA: Harvard University Press.

Sarasin, Philipp. 2001. *Reizbare Maschinen: Eine Geschichte des Körpers 1765–1914*. Frankfurt am Main: Suhrkamp.

Schabas, Margaret. 2005. *The Natural Origins of Economics*. Chicago: The University of Chicago Press.

Schaffer, Simon. 2015. "Les cérémonies de la mesure: repenser l'histoire mondiale des sciences." *Annales: Historie, sciences sociales* 70 (2): 409–35.

Schmelzer, Matthias. 2015. "The Growth Paradigm: History, Hegemony, and the Contested Making of Economic Growthmanship." *Ecological Economics* 118: 262–71.

———. 2016. *The Hegemony of Growth: The OECD and the Making of the Economic Growth Paradigm*. Cambridge: Cambridge University Press.

Schultz, Theodore W. 1962. "Investment in Human Capital." *The American Economic Review* 51 (1): 1–17.

Sewell, William. 2010. "A Strange Career: The Historical Study of Economic Life." *History and Theory* 49: 146–66.

Simmons, Dana. 2015. *Vital Minimum: Need, Science, and Politics in Modern France*. Chicago: The University of Chicago Press.

Smith, Adam. 1776. *Inquiry into the Nature and Causes of the Wealth of Nations*. London: A. Strahan and T. Cadell.

Sombart, Werner. 1928. "Produktivität." *Weltwirtschaftliches Archiv* 28 (1): 1–32.

Speich Chassé, Daniel. 2013. *Die Erfindung des Bruttosozialprodukts: Globale Ungleichheit in der Wissensgeschichte der Ökonomie*. Göttingen: Vandenhoeck & Ruprecht.

Stäheli, Urs. 2013. *Spectacular Speculation: Thrills, the Economy, and Popular Discourse*. Translated by Eric Savoth. Stanford, CA: Stanford University Press.

Stanziani, Alessandro. 2012. *Rules of Exchange: French Capitalism in Comparative Perspective, Eighteenth to Early Twentieth Centuries*. Cambridge: Cambridge University Press.

Steinmetz, George. 1993. *Regulating the Social: The Welfare State and Local Politics in Imperial Germany*. Princeton, NJ: Princeton University Press.

Strathern, Marilyn. 1988. *The Gender of the Gift: Problems with Women and Problems with Society in Melanesia*. Berkeley: University of California Press.

Swanson, Kara W. 2014. *Banking on the Body: The Market in Blood, Milk, and Sperm in Modern America*. Cambridge, MA: Harvard University Press.

Tanner, Jakob. 1999. *Fabrikmahlzeit: Ernährungswissenschaft, Industriearbeit und Volksernährung in der Schweiz 1890–1950*. Zurich: Chronos.

Taylor, Frederick. 1911. *The Principles of Scientific Management*. New York and London: Harper & Brothers.

Tomba, Massimiliano. 2013. *Marx's Temporalities*. Leiden: Brill.

Tooze, Adam. 2001. *Statistics and the German State, 1900–1945: The Making of Modern Economic Knowledge*. Cambridge: Cambridge University Press.

Topalov, Christian. 1994. *Naissance du chômeur 1880–1910*. Paris: Albin Michel.

Trentmann, Frank. 2009. "Materiality in the Future of History: Things, Practices, and Politics. *Journal of British Studies* 48 (2): 283–307.

Tsing, Anna Lowenhaupt. 2005. *Friction: An Ethnography of Global Connection*. Princeton, NJ: Princeton University Press.

Uhl, Karsten. 2014. *Humane Rationalisierung? Die Raumordnung der Fabrik im fordistischen Jahrhundert*. Bielefeld: Transcript.

Vatin, François. 1993. *Le travail, économie et physique, 1780–1830*. Paris: Presses Universitaires de France.

Verheyen, Nina. 2014. "Bürgerliches Leistungsethos? Geschichtswissenschaftliche Korrekturen einer irreführenden Formel." In *Erfolg: Konstellationen und Paradoxien einer gesellschaftlichen Leitorientierung*, edited by Denis Hänzi, Hildegard Matthies, and Dagmar Simon, 45–61. Baden-Baden: Nomos.

Vernon, James. 2007. *Hunger: A Modern History*. Cambridge, MA: Harvard University Press.

Vogl, Joseph. 2014. *The Specter of Capital*. Translated by Joachim Redner and Robert Savage. Stanford, CA: Stanford University Press.

Voss, Lex Heerma van. 2013. "Whither Labour History? Histories of Labor: National and International Perspectives." *International Review of Social History* 58, 97–106.

Vries, Jan de. 2008. *The Industrious Revolution: Consumer Behavior and the Household Economy, 1650 to the Present*. Cambridge: Cambridge University Press.

Wadauer, Sigrid, Thomas Buchner, and Alexander Mejstrik, eds. 2015. *The History of Labour Intermediation: Institutions and Finding Employment in the Nineteenth and Early Twentieth Centuries*. New York: Berghahn.

Weber, Max. (1908) 1995. "Zur Psychophysik der industriellen Arbeit." In *Gesamtausgabe, Abteilung I: Schriften und Reden 1908–1912*, edited by Wolfgang Schluchter, 150–380. Tübingen: Mohr.

———. (1909) 1988. "Debatterede zu den Verhandlungen über die Produktivität der Volkswirtschaft auf der Tagung des Vereins für Sozialpolitik." In *Gesammelte Aufsätze zur Soziologie und Sozialpolitik*, edited by Marianne Weber, 416–22. Tübingen: Mohr.

Wintroub, Michael. 2015. "Translations: Words, Things, Going Native, and Staying True." *The American Historical Review* 120: 1185–1217.

Wunderlich, Frieda. 1926. *Produktivität*. Jena: Gustav Fischer.

Zakim, Michael. 2012. "Producing Capitalism: The Clerk at Work." In *Capitalism Takes Command: The Social Transformation of Nineteenth-Century America*, edited by Michael Zakim and Gary J. Kornblith, 223–47. Chicago: The University of Chicago Press.

Part I

Capitalism and Its Emerging Regimes of Productivity

Introduction to Part I

Mischa Suter and Peter-Paul Bänziger

The global nineteenth century was "a world of connections and inequality" (Cooper 2014; Osterhammel 2014). As the chapters collected in Part I demonstrate, the perspective of body history opens up several lines of inquiry that draw the reader into the global network of industrial capitalism and then beyond. In ways specific to each case, this perspective demonstrates how changing notions of productivity were tied to the body and how particular forms of knowledge mediated these reflections on the productive body. In differing configurations, the body appears as an object of knowledge by means of which concepts of productivity were made effective.

Justus Nipperdey's chapter traces a genealogy of population theory in the early modern period. From the sixteenth century, scholars regarded favorable economic conditions as the prerequisite for a country's population growth and a large population as the prerequisite for its economy's prosperity. Within this reciprocal relationship, some authors placed more emphasis on the economic aspects and others on the demographic ones—but all started from the "extensive" notion that a rising workforce went hand in hand with economic growth. For these theorists, the productive impulse was the reproducing body, which formed a point of intersection between overall economic conditions and the conscious exercise of choice. Seen this way, around 1800, Thomas Robert Malthus, so often regarded as the founding figure of population theory in political economy, stands at the end of a long and tangled chain of thinking on population economics. His pessimistic view of growing populations and scarce resources turned early modern thinking on "bioeconomics" on its head.

Nipperdey's essay demonstrates that bodies were a focal point of early modern economic knowledge. The following two contributions, both situated in West African colonies, are less about knowledge than about non-knowledge: the colonial masters' actively produced and maintained ignorance of the colonized, which helped to forge a "colonial common sense" (Stoler 2009). The gendered body of the colonized formed the core of the "lazy native" myth, as Cassandra Mark-Thiesen explains. Her chapter illustrates how European concepts of productivity took shape within the colonial encounter. These ideas were not built simply upon the notion of an undifferentiated colonial "other", but rather through conflicting stereotypes that might be cast into doubt, yet ultimately reinforced each other. The trope of the male "lazy native" was complemented by the counter-stereotype of the industrious African woman. Here, using gender as a category of critical analysis reveals that ignorance about African work patterns was not a "blind spot", but an active articulation of stereotypes. In the colonizers' eyes, European notions of productivity appeared unrealizable. It seemed impossible to classify the work of African women as productive according to the European conceptual norm for two reasons: first, because the colonizers assumed that value-producing labor could only occur under European compulsion and instruction and second, because women's activities stood outside the parameters of European ideas on productive labor.

Tristan Oestermann's chapter on rubber extraction in German Cameroon investigates the ways in which this "colonial common sense" was underpinned by coercion and violence. The colonizers pitted their allegedly rational extraction methods against African "overexploitation". Yet rubber production increased, not because of, but in spite of colonial control: colonial rubber traders co-opted an existing socioeconomic system without recognizing it as such. Local laborers had an interest in exchanging rubber to earn special-purpose monies, especially for marriage payments, and gave preference to this trade over work as porters—a situation that further fueled colonizers' complaints about the dearth of labor and the "lazy native", culminating in the forced recruitment of porters. The bodies of the colonized, in other words, were the site of a violent intersection between colonial reason and divergent economic regimes, along with attempts to control production.

If Oestermann and Mark-Thiesen discuss the productive body in a colonial context, Sandra Maß addresses a hitherto neglected locus of capitalist subject formation in the metropolis: the monetary education of children, a case that exposes the precariousness of attempts to fashion an economic habitus. The demands of economic education were built primarily on moral and religious discourses. In the early nineteenth century, the educationists who were popularizing knowledge regarding political economy were concerned with moderating the passions (including the biblical sins of avarice and profligacy), but toward the end of the century, the source of moral orientation became the notion of conscious consumption (Nina Mackert's essay in part II illuminates this from a different perspective). Games such as play shops, debates over pocket money, and narratives of the street as a site of work and play for the children of the lower classes suggest that bodily control and the embodiment of behavior were central to the formation of capitalist practices in a process that was convoluted, always contradictory, and frequently obstructed.

In a very different domain, François Jarrige also tracks the "unevenness" (Harootunian 2000) of capitalist modernity in Europe. Jarrige's contribution examines the gradual mechanization of the French bakery trade in the nineteenth century. In France, even today, the majority of bakeries are small businesses. This longevity of an artisanal form of production can be explained by the cultural coding of body practices. Bakers' bodily, tacit knowledge was ascribed great significance by their contemporaries. Their ability to knead dough with hands or feet, the heat of their bodies, and their sensory perceptions were all considered crucial to French bread. The reclassification of the baker's body was driven not only by economic arguments alone but also by the hygienic worries that began to gain ground in the second half of the nineteenth century—due not least to the advent of consumer associations. In parallel and counterpoint to this gradual change, a new and abstract concept of productivity evolved that would develop new facets in the "productivism" of the twentieth century analyzed in part II of the present volume (Rabinbach 1992).

Bibliography

Cooper, Frederick. 2014. "A World of Connections and Inequality: A New History of the Nineteenth Century." [Review of Osterhammel 2014] September 1 2014. Accessed February 24, 2016. www.publicbooks.org.

Harootunian, Harry. 2000. *History's Disquiet: Modernity, Cultural Practice, and Questions of Everyday Life*. New York: Columbia UP 2000.

Osterhammel, Jürgen (2009) 2014. *The Transformation of the World: A Global History of the Nineteenth Century*. Translated by Patrick Camiller. Princeton: Princeton UP.

Rabinbach, Anson. 1992. *The Human Motor: Energy, Fatigue, and the Origins of Modernity*, Berkeley, CA: University of California Press.

Stoler, Ann Laura. 2009. *Along the Archival Grain: Epistemic Anxieties and Colonial Common Sense*. Princeton: Princeton UP.

2 Transgressing Static Concepts

Population, Economy, and Growth in Early Modern Bioeconomics

Justus Nipperdey

While reflecting "the causes of greatness and magnificence of cities", the Italian political thinker Giovanni Botero (1544–1617) pondered the significance of bodily reproduction for the overall growth of a city's population.[1] Although human reproduction was the basis of its preservation or even possible growth, it was by no means the sole determining factor. Botero ([1588] 2012, 73) supposed, "that cities grow in part because of the procreative power of human beings, and in part because of the cities' power to sustain them". In analytical terms, the former did not take center stage:

> It is beyond doubt that the procreative force has remained constant, at least for the last three thousand years, because human beings today are just as capable of reproducing as they were in the time of Moses or David, so that if there were no other obstacle, human beings would have propagated endlessly, and cities would have grown infinitely.

Thus, for Botero, the failure to grow continuously could not be attributed to biological factors. In fact, staying constant over time, the natural "procreative force", and its constituting corporeal elements, such as insemination, pregnancy, and birth, did not hold much interest for the theorist of population development.

Instead, Botero concentrated on the other force, which he called "virtus nutritiva"—the power to sustain people. This could manifest itself in a literal form, designating the amount of food available for the population of the world or of a single city. At some point, the troubles and costs of obtaining food for large cities could no longer sustain population growth. For Botero, however, *virtus nutritiva* had figurative meaning beyond questions of food supply and starvation. It designated the individual's possibilities for earning a living in a certain city that decisively influenced that city's potential for growth. A city that did not offer such opportunities prompted its citizens to leave and look for a better life elsewhere. Moreover, even if people stayed at home, "poverty and scarcity deprive many of the desire and the means to marry and have children" (34), thus generating a negative demographic effect that was not as easily perceptible as emigration. Whereas the natural procreative ability and potential was constant, the economic and social conditions for reproduction were volatile. They tended to shape the demographic development of a society far more than the biological factors.

These insights into the dynamics of both local and global population development constitute an important point of departure and of reference for early modern population thought (see Descendre 2009, 153–71; Nipperdey 2012, 65–98). Taking their cue from Botero, many European thinkers of the seventeenth and eighteenth centuries posited a close relation between demographic development and economic performance of a city or commonwealth. This connection was perceived to be reciprocal. That is, not only would a thriving economy lead to population growth, but a burgeoning population would instigate further economic expansion by consuming and producing goods. Ultimately, this meant that population increase and economic activity were perceived as a positive feedback loop, stimulating each other and leading to overall growth—in people and riches. One issue remained open for debate among the adherents of this theory: Was it possible to set in motion this virtuous cycle from only one side or from both sides of the equation, namely, either by promoting economic activity so as to enlarge the population or by prompting demographic growth by way of population policies to foster economic activity?

In describing this early modern model of connecting population and economy, this chapter consciously uses modern terms such as economic performance, economic growth, or demographic growth. In my opinion, this usage is justified by the fact that the (early) modern idea of growth was explicitly formulated in this field. Furthermore, in combining economic and population thought, this chapter makes use of Catherine Gallagher's concept of "bioeconomics", which forms the basis of her 2006 book *The Body Economic*. Gallagher (2006, 3) defines bioeconomics as referring

> to political economy's concentration on the interconnections among populations, the food supply, modes of production and exchange, and their impact on life forms generally. In bioeconomics, "Life" is both the ultimate desideratum and the energy or force that circulates through organic and inorganic nature.

Whereas Gallagher locates the foundation of bioeconomics in Thomas Robert Malthus's seminal *Essay on Population* (1798), this chapter claims the existence of a much longer tradition going back to the sixteenth century. The formation of bioeconomics in that century is epitomized in the work of Giovanni Botero. Rather than placing Malthus at the inception of the discourse as Gallagher (1986, 83–106) has argued, this contribution presents him as instigating a major turnaround by placing a notion of productivity instead of growth at the center of bioeconomic deliberations. In early modern bioeconomics, attention was directed at the multiplication of workers and not at their individual output. This changed fundamentally with

Malthus's momentous contention that more workers did not equal greater prosperity.

Still, there are obvious connections between Botero and Malthus, as both asked similar questions regarding the relationship of economics and population development. There is a certain irony in the fact that Botero has long been perceived as a direct precursor of Malthus because of his explicit deliberations on the limits of population growth.[2] Reversing this assessment, recent research has placed the Italian thinker at the helm of early modern populationism (Senellart 1989, 56–83; Descendre 2009, 157–70). Rather than preaching abstinence to limit population growth like the English clergyman, the former Jesuit favored population growth. This stance toward population must be understood as part of a larger conceptualization of politics and statecraft that Botero developed in his *Reason of State* (1589) and other writings. In trying to counter Machiavelli's vision of statecraft, Botero put strong emphasis on the internal improvement and strengthening of the princely state. His influential version of reason of state insisted on economic and population policy as a means to safeguard the state and the ruler against overzealous neighbors. On this basis, the concepts of willfully enlarging the population and of stimulating trade and manufacture entered the mainstream political discourse (see Charbit 2010, 63–9). Thus, from Botero onward, reason of state and bioeconomics were deeply enmeshed.

The following chapter will present the bioeconomic discourse between Botero and Malthus in more detail. Early modern bioeconomics heavily focused on the growth of men and things. The central theme of human procreation was considered to be a prime object of political thought and action for two reasons: propagation was perceived as providing for growth in economic production while concomitantly being governed by the economic situation and, thus, serving as a means to measure the economic situation. As the first section will demonstrate, the population theorists changed the role and significance assigned to the reproductive body in order to establish the economic conditionality of population development. An early modern bioeconomic discourse advancing a model of growth could only be established on this basis. The model's overall development and distinctive national modes are analyzed in section two. The concluding remarks will address the striking differences between early modern and later versions of conceiving productivity. Ultimately, the Malthusian critique of the veneration of population growth and the assertion of a modern concept of productivity in Smithian economics displaced the early modern bioeconomic model.

The Body in Reproduction

Giovanni Botero's claim that the natural procreative abilities of men and women were constant both over time and in different geographic locations was far from undisputed. On the contrary, most early modern commentators presumed significant discrepancies between peoples because of corporeal variations induced by geographic and climatic factors, or by these peoples' respective moral conduct. Moreover, when querying whether the rate of procreation was governed by biological or by economic forces, Botero called into question the significance of the human body as an agent of reproduction. This contradicted the patent visibility of the bodies' role in reproduction that was self-evident to people at the time. Research on pregnancy, childbirth, and the female body has shown "the ubiquity of the pregnant woman in the early modern consciousness" (Rublack 1996, 85–6). However, by relegating this aspect to a constant and, therefore, theoretically negligible element, Botero initiated a discourse on population dynamics that curiously ignored exactly this corporeal aspect. It dispensed with all the intricacies of individual insemination and birth that were so widely discussed in early modern culture (see McTavish 2013). Moreover, by emphasizing economic causes over corporeal factors, the bioeconomic argument sharply breaks away from those kinds of connections of "body and politics" that Mary E. Fissell (2004, 2) analyzed in *Vernacular Bodies*. All in all, the bioeconomic disregard of biology and the role of the human body in reproduction seems rather unusual. It is, therefore, useful to briefly introduce some of the prevalent notions that Botero and others had to tackle when arguing that reproductive power was biologically constant and that procreative behavior was determined by economics.

Contrary to Botero's conviction of constant reproductive powers, most early modern authors believed in their diversity and volatility. This issue can be split up into two fairly unrelated aspects, basically differentiating between time and space as reasons for diversity. On the aspect of change over time, early modern natural philosophers and theologians kept grappling with the accounts of the earth's rapid peopling after the Fall and the Deluge, as well as with the proverbial biblical age of the patriarchs (see Egerton 1966, 2005). Both obviously contradicted the contemporary capacities of mankind and suggested an alterability of fertility and longevity. Botero did acknowledge this problem when referring to Moses and David and stating that human procreative ability had been constant for at least the last three thousand years. He echoed the prevalent resolution that posited a connection between the patriarchs' longevity and an analogous prolonged female fertility. According to this notion, God had engineered both an enhanced fertility to people the earth and its subsequent reduction when the world was finally filled. In effect, a constant, more or less natural fertility during post-biblical human history could be reasonably assumed. This conviction was, however, confined to erudite debates on human reproduction and the Bible. Moreover, it only solved the temporal problem.

In contrast, early modern histories, cosmographies, and political pamphlets abounded with comparisons of more and less fertile people. According to these narratives, prowess or deficiency in fertility could have different reasons, ranging from the natural environment to peoples' customs. Most often, these factors were thought to directly influence the human body—be it by reason of climatically induced sterility or illicit sexual customs that reduced procreative capacity. The most common explanation for differing fertility was climate. By calling the European North the "officina gentium" or "vagina nationes", the sixth century historiographer Jordanes (1655, 613) had initiated the topical representation of the north as especially fertile. In the seventeenth century, the Swedish scholar Olaus Rudbeck (1679, 46) expounded on the idea, claiming that Swedish women were usually having between eight and sixteen babies, with the number rising to "24, 30, & 34" in special cases. The other example of procreative preeminence was France. French claims to this status went back to the late Middle Ages. In the early modern period, they were integrated into the emerging populationist discourse (Spengler 1942, 9). Even foreigners such as the English writer John Evelyn observed the "prolifique multiplying" of the French (Tuttle 2010, 27). French human fertility was constantly linked to the fertility of the land and mostly explained by the moderate climate. Whereas Jean Bodin's ([1566] 1969, 85–95) famous climate theory did not explicitly feature procreative ability, his lauding of the temperate zone and its positive impact on the physical and mental equilibrium of its inhabitants clearly corroborated the notion of French fertility.

Conversely, Spain stood on the other end of the spectrum of perceived fertility. The country's lack of an adequate population size was one of the constant themes of early modern population thought. Authors specified a host of reasons ranging from the loss of men through continuous warfare all over Europe and the colonization of America, to the barren soil, or even to the conduct of the Inquisition that, according to Protestant authors, deterred immigrants. In addition, they posited a peculiar sterility closely linked to the bodily disposition of Spanish women and the sexual conduct of Spanish men. Having enumerated the aforementioned reasons, the French travel writer Claude Jordan de Colombier (1704, 14) laconically concluded, "And we must add to all this, the sterility of Spanish women who do not have as many children as those in the Northern countries". Already in the sixteenth century, the cosmographer Sebastian Münster (1544, 46) had compared them unfavorably to the French. "The Spaniards' complexion is warm and dry, and their color brownish; but the French complexion is cold, moist, and whitish. In giving birth, the French women are more fertile than the Spanish". Indeed, these concerns were echoed in the Spanish debate on depopulation beginning in the late sixteenth century. The *arbitrista* Martín González de Cellorigo (1600, fol. 17v.) declared that Spain's sterility "is not a defect of the land but rather a failure of the people. Because the land is quite capable of producing as many people as needed for civil life and for sustaining more than it does".[3]

These exemplary voices are referenced out of a myriad of similar ones so as to illustrate the broad and powerful early modern discourse on the interlocking effects of climate, body, and customs on procreation. The discourse's basic assumption that the procreative power of peoples was as diverse as that of individuals could always be supported by the use of ancient sources. In addition, from the sixteenth century onward, the reports on local customs pouring into Europe from all over the world were seen to affirm this ancient view. It was against this background that Botero developed his generic theory of procreation.[4] It contradicted both the narratives of national or regional diversity and the abounding anxieties of early modern men and women concerning individual fertility.

In omitting these views, the Italian theorist acted strategically so as to fortify the first part of his principal argument: that neither some individuals' lack of fertility nor some conceivable differences between peoples played a determining role in the development of a population as a whole. In his view, procreation was not controlled by the capacities of the body, but by the deliberate choice of the mind, with economic considerations or possibilities guiding this decision. Uncoupling reproduction from the corporeal capacities by declaring them constant and irrelevant was a prerequisite for putting forward the economic argument and, thus, establishing the basis for "bioeconomics". A central element of bioeconomics was its focus on the totality either of the population or the economy. In Botero's theoretical gaze at populations as a whole, all individual circumstances evened out into an overall average, even if he did not try to quantify this average as later generations would attempt to do. Inspired and informed by Botero's considerations, the ensuing population discourse of the seventeenth century followed this line of reasoning. It further developed bioeconomics, placing the interaction of economy and population at the center, while passing over most of the contemporary concerns about procreation and the body.

Early Modern Bioeconomics: A Model for Growth

The assumption of a reciprocity between population and economy lies at the heart of early modern bioeconomics. This chapter's introductory remarks have demonstrated how Botero conceptualized the economic impact on demography by introducing the concept of *virtus nutritiva*. Despite some claims that the early modern populationist thinkers did not reckon with any limits on population growth, they were well aware that a large population needed to be fed and employed if it were to be useful to the prince or the state. Johann Joachim Becher (1688, 2), the most important German seventeenth-century thinker on population, made this point very clear:

> It is, however, not enough to make a city or land populous, if the sustenance is missing; as for a populous congregation to persist, it has to have something to live, aye the latter is the origin of the former.

For him, this observation discredited direct population policy measures, such as penalties for the unmarried, tax incentives for having children, special inducements for immigrants, or religious liberties denied elsewhere.

Many other bioeconomic thinkers deemed such measures nonsustainable as well because they allegedly missed the root of the population problem. To them, both procreation and migration were first and foremost guided by economics. Veit Ludwig von Seckendorff (1703, 164) argued in this vein when he disparaged the argument that immigration to the Netherlands—the very symbol of a populous and rich commonwealth in early modern Europe—was caused by the Republic's religious tolerance. The only reason why people went there, he quipped, was "that in said place everyone who is healthy, be they young or old, can every single day find something to earn".

The Dutch Republic was used not only to illustrate the positive impact of a thriving economy on population size but also to describe a complementary effect: By virtue of its sheer existence, a large population would generate economic activity and, thus, create its own sustenance. As has been shown, this assumption was the second central element of early modern bioeconomics. Nonetheless, it was less self-evident than the dependency of the population on economic prosperity. Considering the constraints of preindustrial economies and societies and the very real threat of famine and hunger, this argument could even be called counterintuitive. In fact, the refutation of this claim proved to be the strongest argument of the opponents of expansive population politics. They could, and did, reasonably argue that population growth would lead to widespread impoverishment, famines, and mortality crises rather than to overall prosperity (see Nipperdey 2012, 557–63). Hence, contemporary authors, faithful to the reason-of-state argument of the interlocking growth of population, overall wealth, and the prince's power and security, put ample effort into explaining their thesis. While adhering to the same basic model, they did come up with slightly different variations. Again, we can turn to Botero (1589, 174) for an early description of the usefulness of a larger population. In his seminal *Reason of State*, he asked why Italy and France were richer than Spain, although they possessed neither gold nor silver mines nor colonies that would furnish such precious metals. For Botero, the only reason for the greater wealth was the much larger population of those two countries. In addition, he attributed France's natural abundance to its large population and explained that the Spanish soil's infertility was not because of natural barrenness but because of a lack of people, a problem that had been hugely exacerbated by the expulsion of the Jews and Muslims. More inhabitants would work and ameliorate the soil and could, thus in turn, provide for the sustenance of the larger population and even produce a surplus (175).

When contemplating the success of the Dutch Republic in the second half of the seventeenth century, authors began to turn their interest to other positive effects of large populations. According to William Temple's (1673, 186) classic *Observations upon the United Provinces*, Holland had grown rich neither by native commodities nor by positive natural circumstances, "but by force of industry". This observation begged the question "what it is that makes people industrious in one Countrey, and idle in another" (1673). For Temple, the answer was obvious:

> I conceive the true original and ground of Trade, to be great multitude of people crowded into small compass of Land, whereby all things necessary to life become deer, and all men who have possessions, are induced to Parsimony; but those who have none, are forced to industry and labour.

Almost anticipating modern theories of habitus, industriousness, and capitalist spirit, Temple added, "These Customs arise first from necessity, but encrease by Imitation, and grow in time to be habitual in a Countrey" (187). To back up his claim, the English diplomat pointed to Ireland as the perfect counterexample:

> Where by the largeness and plenty of the Soil, and scarcity of People, all things necessary to life are so cheap, that an industrious man, by two days labour, may gain enough to feed him the rest of the week; Which I take to be a very plain ground of the laziness attributed to the people. (188)

Juxtaposing industrious Dutch and lazy Irish was a classic moral argument of the time,[5] but for our purposes, the underlying economic and anthropological concepts are more fundamental. In Temple's example, it is not the inherent moral quality of Dutchmen that separates them from the Irish. Just as in propagation, the capacities of inhabitants of different countries are viewed as equivalent. The differing outcome is explained by the distinct economic and demographic circumstances. The high population density, with its accompanying high cost of living, forces the Dutch workers to become "industrious". In the seventeenth-century sense of the word, this meant both moral rectitude and laboring harder. Accordingly, the alleged Irish laziness signified both the missed opportunity to enrich and improve their commonwealth by extending overall production and a form of moral decay. In the evolving "industriousness discourse", high wages were sometimes viewed as a threat to industriousness, as workers would scale back their efforts when they could secure their livelihood in fewer working hours (Litvine 2014, 563–4). According to the bioeconomic discourse, this danger could be averted by demographic pressure on prices and wages.

The argument that a large population would, by its sheer existence, eradicate idleness enjoyed popularity among economic writers of the time. They routinely blamed the laziness of workers for poverty, unemployment, and every imaginable economic difficulty. A large population was conceived as a way to discipline the populace and to implant industriousness as a positive value. For this purpose, some of the economic writers demanded the establishment of workhouses. Contradicting the common continental practice of deporting beggars from a city or a territory, Johann Joachim Becher (1688,

244) even called for inviting those strangers in and transforming them into substantial and useful citizens in the workhouse he envisioned. Still, workhouses could hardly employ all the idle hands. The pressure exerted by competitive overcrowding seemed to open up a broader way of achieving the intended re-education and customization of the workers. In addition, this solution best suited those authors who believed that coercion was not the best way to regulate economic and procreative behavior. Instead, following Botero's lead, many exponents of early modern bioeconomics favored what Michel Foucault (2007, lectures 3 and 4) has termed a governmental approach—that is, reckoning with and using the subjects' own interests and manipulating them in such a way so as to obtain the government's goal instead of trying to attain this aim by force.

In theory, the reciprocal reinvigoration of population and economy opened up multiple starting points for political intervention: fostering either economic or demographic growth, or both at the same time. In practice, opinions varied widely concerning the feasibility and suitability of certain political measures. Notably, direct population politics were controversial among the bioeconomic authors. Whereas almost all of them welcomed skilled immigration, they differed in opinion regarding the necessity of directly encouraging the resident population to have more babies. Rejection—or simple neglect—of this issue was most prominent in the German discourse.[6] There early seventeenth-century authors, including many university professors teaching politics, had embraced Botero's arguments. Whereas some professors repudiated the underlying claim of a necessity of population growth and, consequently, rejected population politics altogether, the majority adopted both the responsibility of a ruler to channel the development of his population and the basic thesis of the desirability of population growth. Moreover, this group took up Botero's economic reasoning, albeit rather tentatively. In their discourse, it was relegated from a central place to being just one factor affecting population development. Conversely, authors such as Hippolyt von Colli (1600), whose *Incrementa urbium. Sive de causis magnitudinis urbium* clearly echoed Botero's own title, tried to assemble a complete list of all factors weighing in on a population without prioritizing any single one. In addition, he and his colleagues collected a multitude of instances of ancient population policy measures, such as the Roman Lex Papia Poppaea, which imposed penalties on celibates. Consequently, this literature contains a host of examples of enforcing marriage and of shaming or fining the unmarried, actions which were presented as wise policy regardless of their suitability to early modern society.

Half a century later, however, a new generation of German-writing authors dismissed all these measures as both futile and intellectually unsound. Wholeheartedly embracing a purely economic interpretation of demographic development, they argued it was impossible to force people to procreate if the conditions were not attractive. These authors took the exclusion of any biological or corporeal matters from the population discourse to a new extreme, considering only economic policy, such as the strengthening of manufactures. This, in turn, was supposed to initiate the population growth they envisioned. Matters such as frequency of marriage or birth rates were not taken into account seriously until the mid-eighteenth century (see Nipperdey 2011, 611–36). Correspondingly, in the German lands, neither concrete policy measures nor deliberations of direct biopolitical interventions in favor of population growth are known.[7]

In France and Britain, the situation was different. In those two countries, bioeconomics was conceived less as a mono-directional analysis as it was in Germany. It seemed possible or even desirable to initiate the virtuous cycle of economy and population from the latter side. In fact, it was Louis XIV's powerful minister, Jean-Baptiste Colbert, who was the only leader of the time to actually enact pronatalist policy. Like their German contemporaries, French intellectuals lauded Roman population policy, arguing that Rome's "increase was also promoted by the rigorous laws against cœlibacy, and the priviledges granted to such as had many children; for propagating which, before Christiauitiy (sic), they permitted polygamy and divorce, and authoriz'd concubinage by legitimation of natural issues".[8] As Leslie Tuttle (2010, 41–61) has shown, Colbert and his advisers consciously tried to follow the Roman model, albeit within the constraints of Christianity. Nevertheless, the influence of the Church and the intricacies of French tax law doomed most of the proposals, involuntarily proving the virtual impossibility of adopting ancient laws in the early modern context. In the end, "what remained of the king's pronatalist policy were the carrots rather than the sticks, the promises of privileges and benefits for early marriage and prolific procreation" (54). Tax breaks or royal pensions for the already tax-exempt nobles were granted to fathers of at least ten living children, of whom none had taken a religious vow. Mothers, upon whose health and fertility such prolific childbearing rested, were not even mentioned in the edicts. Ultimately, however, not even these privileges could be maintained, as their implementation led to countless lawsuits. After one and a half decades, the edict was unceremoniously scrapped.

Colbert's policy did acknowledge economic interests and their role in procreation. It differed from the Boterian and German interpretation of the bioeconomic model by focusing on the individual's monetary interest rather than on the overall state of economic affairs. In addition, the high threshold of ten living children meant that, in reality, only the affluent were targeted, as only they could and did reach the required household size. Thus, the demographic impact would have been trifling, even if the law had remained in force. It can, therefore, be argued that it mainly served symbolic objectives. Still, even in its curtailed form, it remains the greatest experiment in pronatalist policy in early modern Europe and testifies to the widespread belief that population growth was desirable and would stimulate economic activity. Viewed from this perspective, it was negligent to wait for any economic prosperity to incite increased procreation. Authors who were not constrained by the reality of legislative procedure, as Colbert had been, drew more radical conclusions from this understanding of the relationship of economics and population. Notably, this is true for English projectors around 1700.

The best-known English population theorist of the late seventeenth century is undoubtedly William Petty. By systematically introducing quantification into economic, political, and social thought, the polymath and founding member of the Royal Society created the method of *political arithmetic*.[9] Petty connected various strands of seventeenth-century population thought, ranging from the erudite reflections on the peopling of the earth, to more mundane matters of the relationship of economy and demographic reproduction. It is not necessary to spell out all of his ideas on population in detail. Suffice it to say that he not only adhered to the widespread bioeconomic persuasion of reciprocity between population and economic prosperity, but that he also vigorously contended that an initial population growth would generate economic growth. He established this connection on the basis of his reflections on land value and rents:

> The addicon of the people doth not onely increase the rent in proporcon to the respect of numbers, but in proportion to the squares of such numbers, as for example, if the inhabitants of anie country bee 3 millions and shall bee increased to 4, the value of the land shall not only increase from 3 to 4, but from 3 times 3, which is 9, to 4 times 4, which is 16.
> (Quoted in Fox 2009, 395)

As Petty viewed the value of land and the rent it yielded as an important indicator of prosperity, it is possible to understand them as a proxy for economic performance. Thus, in an indirect manner, he posited geometric economic growth to be triggered by arithmetic population increase—a calculation that Malthus would turn upside down a century later. In addition, he argued that a larger population would lead to increased urbanization, which would facilitate production and stimulate consumption, "as the Inhabitants of Cities and Towns, spend more [on] Commodities, and make greater consumptions, than those who live in wild thin peopled Countries" (Petty 1690, 73). All in all, Petty's reckoning suggests an enhanced productivity per person, which was rare at the time. This is a theory that this chapter's concluding section will revisit.

Petty advocated population growth by any means and deliberated the most expedient ways to attain that goal. He directed his fantasies of social engineering at Ireland, proposing either a massive increase of the Irish population or its almost total transplantation to England. In the latter case, England would benefit by gaining one million inhabitants and five million pounds per year, "leaving in Ireland onely enough Hands to manage as many Cattle as that Countrey will feed" (Petty 1899 2: 551). This infamous proposal followed his parallel reflections on systematically enlarging the Irish population from 1.25 to 4.5 million in the space of a single generation. This project, too, would necessitate major population transfer, this time of English "adolescents and women of childbearing age" (Fox 2009, 397). In this context, Petty penned a two-page note "Of Marriages &c." that contained several of his most radical biopolitical conceptions (Petty 1927, 50–1). In essence, Petty proposed an altered form of marriage, which he simply designated as "covenant" and which was supposed to expire once Ireland had reached a population of five million. He appropriately called it "short marriage" for "this covenant may bee dissolved in 6 months, in case of no conception". Furthermore, it "shall produce a child once in 3 yeares whilst they are teeming, under a penalty". Additional provisions to increase procreation included payments "for a woman lying in" and for raising the child to the age of seven, as well as absolving women from the obligation to "declare the father" and issuing rewards for "women that can produce 3 children in 3 yeares" or more than ten children in an unspecified period. These proposals were far from being a broadly shared opinion, or a politically or administratively feasible undertaking. Nevertheless, they do highlight the implications of the bioceonomic model, if taken to its logical conclusion.

In sum, while positing a unitary and widely accepted assertion of the connection of economics and population, the early modern model of bioeconomics offered very diverse ways of thinking about fertility and population politics. In a peculiar way, these different versions unfold in their respective approach to the human body as an object of intervention. For those theorists who focused solely on the economic side of the equation, the body was a *quantité négligeable*. None of the corporeal elements, such as fertility, insemination, and possibly even birth control, shaped the demographic process in any meaningful way. In contrast, for those who sought to initiate the positive process from the demographic side, manipulating procreation became a necessity. They, too, showed little interest in the complicated physical and medical features that fascinated most of their contemporaries. Rather, true to their basic bioeconomic conception, they viewed procreation as a matter of incentives. As for Botero, matters of individual fertility would even out in an average of the overall population. So, even though the body became a prime target of intervention, success or failure of such population policies was detached from the vagaries of nature.

Growth and Productivity in Early Modern Bioeconomics

The early modern bioeconomic model offered the promise of growth, maybe even of boundless growth. In a world that was conceived of as rather static, this was a new concept, which was destined to hold sway over modern economic thought and society at large. Hence, its inception must be viewed as the most important long-term contribution of the bioeconomic discourse. The proponents of bioeconomics had much less to say about a concept that has, in the meantime, become inseparably linked to and posited as the basis of growth: productivity. Looking at modern economic theories of growth, this is not surprising. In the most basic form, they posit two sources of economic growth: increasing labor supply (population) and increasing productivity. As the bioeconomic model is, in essence, about population growth and its connection with economic growth, the predominant, albeit not universal, neglect of productivity seems plausible. In addition, the slow pace of technological change and widespread underemployment made it less likely for authors to focus on productivity in the modern sense. If and when they approached productivity, it was in bemoaning the lack of the individual workers' productiveness because of too few working hours. The main concept of increasing an individual's production was, thus, the extensive addition of working hours, rather than the intensification of productivity. On the level of overall economy and society, extensive addition equaled the increase of workers. In this sense, the productive element was the procreating body.

Furthermore, productivity is a genuinely economic concept that only slowly developed in early modern times. In contrast, the essence of the bioeconomic model is that it is not confined to the economic sphere, but combines economics and population thought. Growth, therefore, did not have to be generated inside the economic sphere alone. It was, this chapter posits, exactly this connection of population and economics that allowed early modern authors to theorize growth and to invent the concept of growth in a relatively static economic environment. With notions and measures of overall economic performance still either unknown or—as in political arithmetic—in an infant state of development, it was difficult to conceptualize economic growth (see Murphy 2009). This was very different in the realm of population. Here the possibility and visibility of growth were self-evident, and even its measurability was enhanced during the sixteenth century. From this vantage point, two steps were necessary to transpose the concept of growth to the economic sphere: (1) the reevaluation of a large and growing population as an unequivocal good instead of a danger to the sustenance and nourishing of the individuals and to public order as a whole and (2) the linking of demographic expansion with economic prosperity. Both steps were taken consecutively in the sixteenth century, first in the reason-of-state discourse and, subsequently, in the bioeconomic discourse emanating from that.

Growth as conceptualized in the bioeconomic model rested mainly on the increase in human population. Both demographic and economic growth was, therefore, essentially extensive growth. Here no productivity gains were necessary—but neither were diminishing returns per individual contemplated. Demographic and economic growth were conceived of, for the most part, as linear; that is, every individual added to the population would add the output needed to sustain her or him, which would still enlarge the overall economy and tax returns. This notion of general extensive growth mirrors the thoughts on the individual workers' output. Both were, as we have seen, closely associated in the second half of the seventeenth century, when one of the central benefits of a large population was identified as forcing workers out of their alleged idleness. This campaign against idleness was part of the early modern "industriousness discourse" that has received much scholarly attention in recent years, not the least because of Jan de Vries's concept of an "industrious revolution" as a precursor and prerequisite of the Industrial Revolution (de Vries 2008; Litvine 2014).

Many bioeconomic authors enthusiastically endorsed the propagation of industriousness and especially its progressive reinterpretation as an economic concept. In practice, they demanded forcing laborers to work more days and longer hours. This focus is theoretically equivalent to the enlargement of the labor force in the course of population growth. Thus, the goal of industriousness as propagated by early modern authors was finite: Once everyone works to her/his full capacity, the goal is achieved and further growth becomes impossible. In theory, this is supposed to happen rather quickly, so its fulfillment will precede the mid- or long-term growth that is to be generated by population growth. In the bioeconomic discourse, the imposition of industriousness can be understood as a necessary first step to make use of all human potential that is to be expanded by procreation in a second step.

Notwithstanding this general trend, some bioeconomic authors did allude or even clearly refer to growing productivity. Giovanni Botero's remarks about growing agricultural output in case of intensified cultivation can be interpreted in this way. The most articulate proponent of increasing productivity in the bioeconomic cycle was again William Petty. In contrasting the arithmetic growth of the population with the ensuing geometric progression of land value and overall wealth, he clearly spelled out a concept of rising productivity because of the larger population. In addition, his remarks regarding enhanced production in large cities point to his belief in a more productive use of human labor in the densely populated state that he envisioned. Still, for most bioeconomic authors, productivity rested in the human body and its procreative potential. Despite their general disregard for the particulars of fertility and procreation, the engine of the bioeconomic mechanism consisted of humans producing more humans. Productivity was centered on procreation.

This dominant notion of the seventeenth century slowly eroded during the age of Enlightenment, only to receive its deathblow at the hands of Thomas Robert Malthus. Whereas productivity remained implicit and unnamed even in the writings of William Petty, the concept and term gained prominence in the eighteenth century. The most common usage, however, referred to the division between productive and unproductive labor. This distinction lay at the core of the physiocratic conception of the economic world and also featured prominently in Adam Smith's *Wealth of Nations* (see the introduction to this volume; Steiner 1992, 232–4; Aspromourgos

2013, 275–8). In addition, Smith famously portrayed the division of labor as the central tenet of rising productivity (Aspromourgos 2013, 268–70). Following this, economic growth could potentially be disassociated from population growth. Still, the central idea of early modern bioeconomics linking the individual body, the population, and the economy was not challenged until Malthus, as Gallagher (1986, 83) has shown:

> With the publication of Thomas Malthus's first *Essay on the Principles of Population* in 1798, the social and economic significance of the vigorous body was radically reconceptualized. By insisting that healthy bodies eventually generate a feeble social organism, Malthus departed from nearly all his contemporaries.

From a Malthusian perspective, generating ever more human beings was not the way to growth and prosperity, but to disaster and misery. Population growth and economic expansion were uncoupled. In pointing to the relationship between economy and population, Malthus did not invent bioeconomics but, rather, radically reconfigured it by turning the earlier model upside down. Concerning the concepts discussed in this chapter, it is interesting to note that Malthus did not further develop the concept of productivity. In his pessimistic view, there was no room for its inherent possibilities of steadily intensifying production and outpacing population growth. While famously allowing for arithmetic growth of foodstuffs, Malthus, rather, reflected the earlier model of extensive growth. Having deprived it of the long-term prospects of population growth, Malthus had, at the same time, destroyed the justification for a belief in constant future growth. It would take the gradual advancement of the modern concept of productivity to rebuild such an optimistic worldview.

Notes

1 *Delle Cause della Gradezza delle Città* was first published in Rome in 1588. From the second edition of *Della Ragion di Stato* (1589) onward, it was attached to that work. The following uses the recent first modern English edition (Botero 2012).

2 See the introduction to a modern edition of Book III of Botero's *Delle Cause della Gradezza delle Città* in the *Population and Development Review* (1985).

3 English translation cited from Jon Cowans (2003, 135–6).

4 It must be noted that Botero did, in fact, take up the climate theory in his *Relazioni universali* (1591–1596), although its centrality remains contested among historians. For differing opinions, see John H. Elliott (2009, 201) and Romain Descendre (2009, 298).

5 Regarding the comparable myth of the "lazy African", see the chapters by Cassandra Mark-Thiesen and Tristan Oestermann in this volume.

6 For more detail and references, see Justus Nipperdey (2012, 212–42).

7 There were, however, ample measures to prevent population increase by hindering marriages of the poor.

8 The question was discussed in the intellectual circle of Théophraste Renaudot in the 1630s. The proceedings were later published in English too (*General Collection of Discourses of the Virtuosi of France* 1664, 451).

9 See his posthumously published *Political Arithmetick* (Petty 1690). For the burgeoning literature on Petty, see Ted McCormick (2009).

Bibliography

Aspromourgos, Tony. 2013. "Adam Smith on Labour and Capital." In *The Oxford Handbook of Adam Smith*, edited by Christopher J. Berry, 267–89. Oxford: Oxford University Press.

Becher, Johann Joachim. 1688. *Politische Discurs von den eigentlichen Ursachen/des Auff- und Abnehmens der Städt/ Länder und Republicken*. Frankfurt am Main: Zunner.

Bodin, Jean. (1566) 1969. *Method for the Easy Comprehension of History*. New York: Norton.Botero, Giovanni. (1588) 2012. *On the Causes of the Greatness and Magnificence of Cities, 1588*. Translated by Geoffrey Symcox. Toronto: University of Toronto Press.

———. 1589. *Della Ragion di Stato libri dieci*. Ferrara: Baldini.

Cellorigo, Martín González de. 1600. *Memorial de la politica necessaria*. Valladolid: Bostillo.

Charbit, Yves. 2010. *The Classical Foundation of Population Thought: From Plato to Quesnay*. Dordrecht et al.: Springer.

Colli, Hippolyt v. 1600. *Incrementa urbium: Sive de causis magnitudinis urbium, liber unus*. Lübeck: Albertus.

Colombier, Claude Jordan de. 1704. *Voyages historiques de l'Europe*. Brussels: de Grieck.

Cowans, Jon, ed. 2003. *Early Modern Spain: A Documentary History*. Philadelphia: University of Pennsylvania Press.

Descendre, Romain. 2009. *L'État du monde: Giovanni Botero entre raison d'État et géopolitique*. Geneva: Droz.

De Vries, Jan. 2008. *The Industrious Revolution: Consumer Behavior and the Household Economy, 1650 to the Present*. Cambridge: Cambridge University Press.

Egerton, Frank N. 1966. "The Longevity of the Patriarchs: A Topic in the History of Demography." *Journal of the History of Ideas* 27: 575–84.

———. 2005. "The Precocious Origins of Human and Animal Demography and Statistics in the 1600s." *Bulletin of the Ecological Society of America* 86: 32–8.

Elliott, John H. 2009. *Spain, Europe and the Wider World, 1500–1800*. New Haven: Yale University Press.

Fissell, Mary E. 2004. *Vernacular Bodies: The Politics of Reproduction in Early Modern England*. Oxford: Oxford University Press.

Foucault, Michel. 2007. *Security, Territory, Population: Lectures at the Collège de France, 1977–1978*, edited by Michel Senellart. Basingstoke: Palgrave Macmillan.

Fox, Adam. 2009. "Sir William Petty, Ireland, and the Making of a Political Economic, 1653–1687." *The Economic History Review* 62: 388–404.

Gallagher, Catherine. 1986. "The Body Versus the Social Body in the Works of Thomas Malthus and Henry Mayhew." *Representations* 14: 83–106.

———. 2006. *The Body Economic: Life, Death, and Sensation in Political Economy and the Victorian Novel*. Princeton: Princeton University Press.

General Collection of Discourses of the Virtuosi of France. 1664. London: Dring and Starkey.
Iornandes [Jordanes]. 1655. "De Getarum sive Gothorum origine et rebus gestis." In *Historia Gothorum, Vandalorum, et Langobardorum*, edited by Hugo Grotius, 607–703. Amsterdam: Elzevier.
Litvine, Alexis. 2014. "The Industrious Revolution, the Industriousness Discourse and the Development of Modern Economies." *The Historical Journal* 57: 531–70.
McCormick, Ted. 2009. *William Petty and the Ambitions of Political Arithmetic*. Oxford: Oxford University Press.
McTavish, Lianna. 2013. "Reproduction, c. 1500–1750." In *The Routledge History of Sex and the Body: 1500 to the Present*, edited by Sarah Toulalan and Kate Fisher, 351–71. Abingdon: Routledge.
Münster, Sebastian. 1544. *Cosmographia*. Basel: Petri.
Murphy, Antoin E. 2009. *The Genesis of Macroeconomics: New Ideas from Sir William Petty to Henry Thornton*. Oxford: Oxford University Press.
Nipperdey, Justus. 2011. "Johann Peter Süßmilch: From Divine Law to Human Intervention." *Population-E* 66: 611–36.
———. 2012. *Die Erfindung der Bevölkerungspolitik: Staat, politische Theorie und Population in der Frühen Neuzeit*. Göttingen: Vandenhoeck & Ruprecht.
Petty, William. 1690. *Political Arithmetick*. London: Clavel.
———. 1899. "A Treatise of Ireland (1687)." In *The Economic Writings of Sir William Petty*. Vol. 2, edited by Charles H. Hull, 545–621. Cambridge: CUP.
———. 1927. "Of Marriages &c." In *The Petty Papers: Some Unpublished Writings of Sir William Petty*, vol. 2, edited by Henry Lansdowne. London: Constable.
Population and Development Review. 1985. "Giovanni Botero on the Forces Governing Population Growth." 11: 335–40.
Rublack, Ulinka. 1996. "Pregnancy, Child Birth and the Female Body in Early Modern Germany." *Past & Present* 150: 84–110.
Rudbeck, Olaus. 1679. *Atlantica*. Uppsala: Curio.Seckendorff, Veit Ludwig von. 1703. *Additiones Oder Zugaben und Erleuterungen Zu dem Tractat des Teutschen Fürsten-Stats*. Frankfurt am Main and Leipzig: Meyer.
Senellart, Michel. 1989. *Machiavélisme et raison d'état: XIIe—XVIIIe siècle*. Paris: Presses Universitaire de France.
Spengler, Joseph J. 1942. *French Predecessors of Malthus: A Study in Eighteenth-Century Wage and Population Theory*. Durham, NC: Duke University Press.
Steiner, Philippe. 1992. "L'Économie politique du royaume agricole: François Quesnay." In *Nouvelle Histoire de la pensée économique, vol. 1: Des scolastiques aux classiques*, edited by Alain Béraud and Gilbert Faccarello, 225–53. Paris: Découvertes.
Temple, William. 1673. *Observations upon the United Provinces of the Netherlands*. London: Gellibrand.
Tuttle, Leslie. 2010. *Conceiving the Old Regime: Pronatalism and the Politics of Reproduction in Early Modern France*. Oxford: Oxford University Press.

3 African Women and the "Lazy African" Myth in Nineteenth-Century West Africa

Cassandra Mark-Thiesen

A history of productivity in nineteenth-century West Africa written from a global perspective needs to consider that over several centuries, Europeans made little effort to measure productivity on the continent in a scientific manner. Indeed, before attempts were made at integrating economic activity in Sub-Saharan Africa into a standardized analytical framework (informed by Western conceptualizations of economic modernity), the tendency was to marginalize, belittle, if not erase altogether, trade and industry in that part of the world. It would take the watershed moment of African independence starting in the 1950s to convince a broader audience of the "indigenous economic dynamism" (Austin and Broadberry 2014, 893) of African societies. There was a long and potent history of arguing against such a perspective.

European interest in African work ethics predating the age of imperialism was part and parcel of a preoccupation with opening up the continent for European influence, exploitation, or control. And as a result, discourses surrounding the issue of productivity here were shaped differently from those taking place in contemporary Europe. Up until the end of the colonial period, when European employers talked about the potential of African workers in various jobs and economic sectors (for instance, their ability to be disciplined to the structure of the eight-hour workday or general steady employment), the ensuing debate often touched on the subjects of racial biological and cultural inferiority, broadly speaking "civilization". The caricature of the "lazy African", made prominent through repeated mention in European travel diaries and official reports starting around the fifteenth century, served to remind European audiences about the lower economic and social status of the African.

The stereotype of the "lazy native" was regularly reconstructed alongside Europe's initial interest and later domination of the African continent, and in this manner, it solidified to the point of presenting a starting point for most observations on African economic behavior, irrespective of whether the evidence supported such a finding. Indeed, many scholars who have worked extensively with European travel diaries or official sources have come to realize that the economic lives of Africans could at times be illustrated with one stroke of a pen and erased in the next. To demonstrate one example, it is poignant to note that the anthropologist Polly Hill (1997, 18) could point to both references of land purchases by Akwapim cocoa farmers in the Gold Coast Colony (Ghana) and complaints about their ignorance about savings in the same official reports. The economic historian Gareth Austin also noticed how the myth continued to be the baseline for recordings of African economic activity in West Africa during the period of colonial rule:

> The insidiousness of that "lazy African" image, even when contradicted by observation, is captured in an 1897 dispatch from H. M. Hull, a British commissioner, who in the very same sentence managed to describe Akims as "not energetic", preferring "doing nothing to" and "keen as petty traders and of course collect large quantities of the Kola nut".[1]

This image, present across travel diaries and children's books alike, was an omnipresent one.

As scholarly research has recently shown, there were certainly nuances to this debate: a) not all Europeans reified this stereotype and b) some even spread contradicting depictions of the African at work. Nevertheless, with regard to the latter point, the question remains as to what extent such observations encouraged more situational (less biological) attributions for the behavior of workers in Africa. This chapter is particularly concerned with contemporary remarks like the following, which contended that idleness "was . . . not ubiquitous in all of African society, but a problem limited to the male population" (Rönnbäck 2014a, 216), supposedly the result of a gendered division of labor that made men's work less visible to European visitors. Consequently, this chapter revisits some recordings made by European men of female industriousness in nineteenth-century West Africa.[2] Without claiming to present a comprehensive overview of the source material, what emerges from these texts is an understanding that acknowledgments of this nature did not necessarily successfully counter the dehumanizing effect of the original myth. Indeed a gender perspective on the myth reveals that the exception formulated in different sources produced by white European men could actually be built upon the same racist foundations as the original stereotype itself, but with an added layer of misogyny.

(Countering) The "Lazy Native" Myth

> While the mind of the traveler is oppressed by the melancholic consideration of the moral and intellectual darkness of the scattered tribes of human beings, whose desolate-looking hamlets frequently meet his view, as he ends his way amidst the dense forests of the uncultivated hills and dales of Africa, he is encouraged to believe that the time will come when this extensive "wilderness shall be made glad" by the labors

> of industrious agriculturists, and when this vast desert of intellectual and moral degradation "shall rejoice, and blossom as the rose".
>
> (J. W. Lugenbeel, Colonial Physician and U.S. Agent in Liberia, c. 1859)

> The Word of God was blessed to many of them. They are all now civilized and married: they are steady, sober, and industrious.
>
> (Observations of the Igbo people of current-day Nigeria as recorded in the proceedings of the Church Missionary Society for West Arica, London, 1819–1820)

Although the evolution of this Western stereotype will be discussed in a strictly African context here, it is important to acknowledge that stereotyping of this nature was not restricted to the African worker.[3] Rather, colonized people all over the globe experienced the same form of racial subordination (Alatas 1977). Poor whites in the metropole faced this type of discrimination, as well (Cox, 261).[4] For Africa, it has been suggested that the spread of the myth even preceded a stage of intensive European travel to the continent (Blackburn 1997, 70). Regardless of how accurate this finding might be, what holds true is that its spread initially intensified during the fifteenth century. From that point on, European travelers and administrators regularly revived the myth to legitimize various economic policies relevant to imperial expansion and colonialism. Broadly speaking, with commerce being at least as significant as "civilization" and Christianity, the "lazy African" myth generally served to justify European intervention (or nonintervention) in Africa at various points in history.

Economic concerns occasionally trumped others, and not only from the state's point of view. Missionaries were also known to underline the importance of industriousness in front of their congregations. For example, when the Swiss zoologist Johann Büttikofer (1890, 74) visited Liberia for a few years starting in the late 1870s, he noted that disciplining supposedly lazy Africans for regular work had become a main objective of a local American Lutheran missionary, Rev. David A. Day[5]:

> His priority is not to impose Christianity onto the Negro at all costs, and also not to turn his pupils into scholars, but rather to encourage them to pursue steady work, to pull them out of their innate laziness and motivate them to engage in productive, profitable work.

Work was a worthy primary target, because a lack of industry was understood to mean much more that just the loss of productive potential. It signaled to some intellectual and spiritual weakness within an individual and or broader population.

Scholars have shown the myth to carry an array of assumptions about the inferiority of Africans at various moments in history: It initially helped to justify the entry of European traders into the transatlantic slave trade. Travelers made Africans out to be lazy, as well as savage, excessively salacious, cannibalistic, and conniving. During the 1830s, this fantasy reinforced the idea of an immutable racial difference with regard to workers of African descent in the British Empire especially. The pseudoscience of phrenology would show Africans to be unsuited for work apart from when they were under European supervision. It aided the acceptance of indigenous slavery in Africa during the Scramble for Africa. The protracted death of indigenous slavery was the result. During the colonial period, "labeling the African worker as lazy was a way of acknowledging the limits of dominance, while attributing these limits to the basic nature of the dominated" (Cooper 1987, 1). Also, as Tristan Oestermann's essay in this volume exemplifies, under particular conditions, colonial officials employed this myth to advocate for the use of forced labor where they desperately wanted to exploit the labor power of African workers. Above all, the undermining of Africans as social, political, and economic actors through this fiction became so pervasive because a wide range of laymen and lobbyists reinforced the myth with varying degrees of malintent. Nevertheless, this chapter is concerned with those travelers and officials who gave a different account of industry in Africa.

The heavily racialized discourse on laziness in Africa encountered resistance as early as the late eighteenth century, as part of an abolitionist discourse. The abolitionist and freed slave Olaudah Equiano was perhaps one of the most prominent individuals to actively restore the tarnished image of African economic life during this era. With the help of anecdotes from his life back in the region of modern-day Nigeria, he shared, among other things, his optimistic outlook on Africa's economic potential with readers who were already interested in the continent:

> Population, the bowels and surface of Africa abound in valuable and useful returns; the hidden treasures of centuries will be brought to light and into circulation . . . In a word, it lays open an endless field of commerce to the British manufactures and merchant adventurer. The manufacturing interest and the general interests are synonymous. The abolition of slavery would be in reality [a] universal good.
>
> (Equiano 1789, 357)

In rational economic terms, Equiano argued for an end to the transatlantic slave trade and the spread of legitimate commerce on the African continent. In other words, he was debunking the myth by arguing in favor of the ability of African workers to create value outside of coercive colonial intervention, thereby appealing to an emerging modern/contemporary notion of productivity most prominently put forward by Adam Smith.[6] Yet, rather than mentioning any concrete measure of African input/output, his appeal rested upon the positive imagery of a vastly populated continent, whose people simply needed to be provided with more opportunities.

During the period that is the subject of this chapter, contradictions to the stereotype were gradually starting to gain traction outside of anti-slavery circles, as well. An ideological shift in the nineteenth century led to an increase in the number of fundamental challenges and contradictions to the myth of the "lazy African". Some European visitors to and residents in Africa were now gaining insight on the work patterns, motivations, and general economic choices of African workers. As Klas Rönnbäck (2014b, 129) has pointed out,

> idleness is . . . not the only description of an African work ethic that can be found in the writings of European travellers in pre-colonial Africa. . . . Quite a few authors argued that Africans, on the contrary, were or could be highly industrious.

Particular "ethnic" groups received acknowledgment for their work ability. African women workers were also often spoken of in contrasting terms to the allegedly idle African man. At the same time, there was significant variance in the claims associated with industrious behavior in different contexts, with some depictions being more successful than others at presenting Africans economic activity away from the racialized biases of the original myth.

West Africa experienced a burst of travel activity in the nineteenth century, likely more than any other part of Africa. During this period, expeditions sponsored by private agencies, such as the African Association,[7] to explore the interior parts of the continent in the eighteenth century were being largely replaced by government-backed voyages with both economic and political objectives (Bridges 2002, 59). The personal interests of the administrators and other visitors involved usually extended beyond a predetermined roster of inquiry to include the topics of geography, taxonomy, and culture. Descriptions of African attitudes toward work often made the list of individual interests as well, most likely because of the broader context in which the subject was situated: "In the minds of many [Western] authors during the early Industrial Revolution, industry meant setting one's mind to the work at hand no matter the difficulty, knowing that rewards for such toil would ultimately be adjudicated by God" (Yeats 2011, 148). Thus, labor was a matter of the present, the future, and the afterlife alike, as seen in the earlier quotes. The last quarter of the nineteenth century was different in that it marked the Scramble for Africa. By this time, colonialism and the rapid spread of capitalism on the continent made the challenge of controlling African labor a priority for European powers. For, in their minds, having control over this labor resource would permit productivity to occur on the continent, which would in turn affect the potential of imperial rule and expansion. However, growing contact between European and Africans did not strictly exacerbate the spread of the "lazy African" myth.

Starting in the nineteenth century, an increasing number of explicit challenges to this stereotype were voiced. Whether they intended to actually reconstruct the image of the African worker or not, these statements at least complicated a salient image that did not seem to adequately capture actual conditions on the ground. Direct and targeted contradictions to the myth provided new situational understandings for the difficulty of mobilizing Africans for various types of steady and extended wage work, a challenge that could be material, cultural, and even spiritual in nature. For instance, when Thomas Joshua Alldridge (1901, 337), the first district commissioner to Sierra Leone was appointed to the Sherbro region in 1894, his observations revealed that it was not laziness but inadequate wages that kept large numbers of African men and women from joining the wage labor market, choosing to work and live in the rural sphere instead:

> What I consider of vital importance to the natives is profitable employment in which all can participate. Oldtime notions have to be eradicated and new ones built up. The easiest and most satisfactory way to do this and to make the people happy and contented is to provide the means of remunerative work. I do not find that people in any civilised country care to work for next to nothing, why then should the untutored African native be expected to do so? But if he does not do so he is too often branded as a lazy and useless creature. A great deal has been said about the scarcity of native labor; but my own experience is that a sufficiency, and more than a sufficiency, can be obtained for the supply of the Sherbro and its Hinterland, when there is a prospect of its being fairly paid for.

Reports like this one helped to turn prevailing prejudices on their heads. Indeed, a greater awareness of the local context occasionally sparked debates among European employers over whether they themselves should adjust to African work patterns, respectively "an African work ethic" (Atkins 1993; Cooper 1996).

Yet, declarations like Alldridge's are far less frequent in the sources. What foreign travelers, missionaries, officials, and the like seemed to have done more often is to make a case for certain exceptions. What Rönnbäck (2014b, 129) calls "the stereotype of the industrious African" would "virtually always be made regarding people of a special ethnicity or nationality in Africa". By the nineteenth century, ethnic identity had become a key instrument in European hiring processes in Africa. Where a worker came from could therefore influence the type of work a European employer deemed him or her to be equipped for, as frequently observed in the hiring tactics of miners or soldiers (e.g., Dumett 1998; Moyd 2014). Ethnicity could influence both the type of work and wage rate offered to a given individual. Deducing information about their ethnicity, employers made a variety of assumptions about the intellect, physicality, and morality of their imagined ideal African worker: What kind of religious background defined a hardworking social group? Where should they fall along the spectrum between "barbaric" and "civilized"? Were the tall and slender more efficent than the short and stout? Did long-term frequent contact with Europeans spoil men for hard labor? Or did it make them more adaptable to wage work? The primary sources dealing with these perceptions show no consensus in any particular direction.

But even if European observers acknowledged and wrote about these ethnic exceptions, to what extent were their words successful at rehabilitating the image of the African worker? The following section analyzes European sources dealing with another group of Africans commonly celebrated for their ability to engage in hard work, namely African women. It compiles a number of descriptions of work ethics in nineteenth-century West African women in order to better comprehend the claims made by white European male observers with regard to women's potential for economic forethought, rationality, industriousness, and productivity, as well as the body of the female African worker.

Gender, Race, and Industriousness

> The women are very industrious and remarkably prolific.
>
> (Gov. Henry Meredith, Gold Coast 1812)

The labor of African women became most important to Europeans toward the nineteenth century. In recent times, according to Diana Jeater (2006, 228),

> few historians would deny that women's labor was crucial in generating cash-crop incomes, in sustaining rural communities and building urban centers, or that the Empire in Africa depended on the preservation of old and the creation of new forms of female subordination, all of which explain why the subject of women commanded so much attention from Colonial Officers.

Yet, as Jeater reemphasizes, the "British Empire affected African women differently than it did African men". Such variation is, for example, seen in the observations and assumptions of nineteenth-century European officials and travelers regarding women's work ethic. The gender aspect of the "lazy African" myth, which has been part of an academic discourse since at least the 1970s, was just recently reexamined (Rönnbäck 2014a).[8] This literature points out that remarks made about African women at work often deviated from the "lazy African" myth, although scholars have made sense of this exception in a number of ways. The development economist Esther Boserup (2007, 20, 55), whose research is frequently mentioned in this context, blamed this bias on European ignorance about African gender division of labor. To be precise, she contended that the work traditionally done by African men (hunting, tree felling, and warfare) became less relevant with the arrival of the European. The colonizers would, as a result, do everything in their power to push the men into agricultural work and food gathering (activities traditionally reserved for the African woman) in increasing numbers because farming was traditionally a male activity in Europe.

In the past, Boserup's critics have chastised her for imposing westernized conceptions, such as a dichotomous gender division of labor, onto work in Africa. She also stands accused of making broad generalizations about economic activity throughout the entire continent and depicting African women in an excessively passive, dependent, and inflexible manner. In addition, the author would add that her aforementioned argument is too quick to except both the objectivity of the vast majority of the original source material and the extent of European influence. In investigating the discursive claims of the counter-stereotype of the industrious female African worker, the following questions seem of greater relevance: What pieces of information circumscribe such an individual? How are these claims put to use? And do they serve to give African women a more human character than the "lazy African" himself?

In rereading sources from the nineteenth century, it is not entirely clear that the observer's "admiration" was necessarily geared toward combating the effect of the original narrative. We cannot ignore the obvious switch from a focus on "idle" to "industry" in Africa. Nevertheless, we also cannot overlook that claims of industry could be highly managed, often with the outcome of still devaluing the economic actors involved. For instance, the dominated, passive, yet, industrious African woman was a figure present in many accounts of European observers. Supposedly lazy male "masters" (i.e., local male chiefs and the men in general) were the ones forcing *their* women to work, and work hard. During an 1841 expedition to the River Niger concerned with ending the transatlantic slave trade and increasing British influence (trade, politics, religion) in the area, Captain William Allen of the British Royal Navy also landed in the vicinity of the Cape Coast (Ghana), the headquarters of the British colonial government at the time and home to an estimated six thousand local Fante people. During his stay, he visited the households of several Fante families, summarizing the division of work between women and men as follows: "the men were found lying about as if there was no possible occasion for working, while the women were employed in some of their simple household duties" (Allen 1841, 2: 135). African men, in other words, were lazy because they could force the women to do all the hard work. While putting lazy men next to women at work, many authors, as will be seen in the course of this chapter, simultaneously invoked the image of African women working in semi-slavery conditions, not so much forcibly subdued as ignorant of their submission.

Allen further noticed that women were also very active in the region's most popular extra-subsistence activity, namely gold mining: "While rambling about the town, we found several women engaged in washing the sand left in the little channels caused by the late rains, from which they procured a small quantity of gold dust". Yet, the captain believed that small-scale gold mining was actually keeping the Fante back from true "progress", because of the speculative nature of the business, "sometimes yielding a fair day's return, at other [times,] nothing". As far as he could tell, however, Fante men were not worried about economic transformation, because the women were the ones investing their labor power, to the men's gain: "as the whole labour falls on the women, the lazy Fante[9] looks on unconcerned, since the expenditure of the proceeds is usually on his own tastes and pleasures—drinking and smoking" (Allen 1841, 2: 136). Thus, European travelers could highlight the presence of African women workers, while still placing them in a category of beasts of burden and representing their work as nonproductive. This was especially easy to do, because in the nineteenth century, women's work largely fell outside of the realm of contract work, even when they did contribute to large-scale commercial production (see, for example, Mark-Thiesen 2014).[10]

The economic activity of hardworking African women was frequently engulfed in gender and racial biases. In the 1870s, while chronicling the Anglo-Ashanti War, the British journalist Frederick Boyle also endeavored to capture aspects of economic life in and around the Gold Coast Colony. Land and labor were of particular interest. It was in that context that he reprimanded the Fante men for not taking greater advantage of the extremely fertile land in their ownership: "Every village has from five to twenty trees planted in it, for the sole purpose of giving shade" (Boyle 1847, 36–7). Fante women were, nonetheless, different. Although they purportedly showed little evidence of order and organization, Boyle claimed to understand quite clearly why transport officers cherished female carriers: "The arrangement, the decency, which had been hopeless to attain with Krumen,[11] or with the villainous [sic] recruits of Sierra Leone, are achieved without difficulty by the women" (30). In making a direct comparison between women workers and their male counterparts, he noted, "in physical force, they are more than a match for those they call their lords".

However, at the same time that he celebrated African women workers, he also claimed that "their tractability seems to be beyond proof", making the point that a degree of irrationality informed their willingness to work. Although he praised the strength, "arrangement", "pleasantness", and the "decency" of female Fante carriers, these African women were, in his view, driven by feeblemindedness. Furthermore, according to Boyle, other European employers were equally well aware of how easily swayed Fante women could be to do such work. In a particularly revealing statement, he recalled being told by a Scottish sergeant that in order to get such women to work, "Just tip ''em a laugh and a punch in the ribs', . . . an' them there [sic!] Fante[12] women will carry 200 pounds each from this to Coomassie on their heads!" Boyle apparently stood by this ludicrous image in his own understanding of why colonized women work, explaining to his readers on the British Isles that the Fante women's "strength is surprising, and their merriment inexhaustible. No work seems to daunt them, if only they get good humoured encouragement from the officers" (Boyle, 30). In this manner, industry was paired with the same childish or foolish characteristics reserved for colonized peoples and women in a variety of contexts.

It is further noteworthy that the concept of womanhood was not transferred onto these productive African women with ease—not that either physical ability or productivity was easily reconciled with femininity, in general. The British journalist described the bodies of these industrious Fante women as "physically androgynous (the counterpoint to the Hottentot Venus)" (41). So authors like him created a counterimage to the symbol of the oversexed (yet, still lesser) African woman. Given the inability to reconcile physical ability and economic achievements with women, masculine traits were projected onto what were now most unfeminine African women.

Another theme that is given repeated attention is the purported dull-mindedness in bodily strong African women. It is usually demonstrated when work is phrased in terms of excessive corporeality. Feminist perspectives on the body have been attuned to the fact that women "are somehow more biological, more corporeal, and more natural than men" (Grosz 1994, 14). Indeed, according to Linda Martin Alcoff (2006, 103), such enmeshment in corporeality was equally attributed to both colonized bodies and those attributed to the lower classes. For instance, in West Africa, women dominated the marketplace in the nineteenth century. While visiting a marketplace in Freetown abundant with commodities such as tropical fruits and "fish, of various kinds", a European traveler first suggested, "nearly all the different articles exposed for sale were proffered by Negro women". In fact, according to him, the women

> did not restrict their commercial transactions to the market, for the 'Wilberforce' was surrounded with busy visitants, each bearing—Ceres-like—some production of the soil, as bananas, oranges, pines, custard-apples, sugar-cane, earth-nuts, etc. They were all liberated Africans, or their descendants, full of noisy mirth and business.

Nevertheless, employing the aforementioned gender tactics, in a later sentence, he denied these market women any astute business sense. Instead, he finally captured their activity with the statement that, "the incessant clatter kept up on every side, quite convinces the visitor that the active use of that little member, the tongue, is not confined to the gentle sex of more temperate climates" (Allen 1841, 2: 80–1). Thus, female labor in this instance was not represented in a quantifiable manner, but rather complied with a patriarchal discourse on production. It was reduced to chattiness, a particular part of the body, namely the mouth, being its lead instrument.

In a personal narrative of their time spent in West Africa in search of gold, the British entrepreneurs Richard Burton and Verney Cameron (1883, 330) expressed in an overt manner that the productivity of African women was not necessarily something to be respected nor admired. In the 1880s, they encountered West African women working on the coast, loading and unloading ships, which was a most unusual site for "men who visit Africa for the first time". They also witnessed female gold miners in colonial Ghana, of whom they related the following information to their audience at home:

> Beyond this settlement is a fine quartz hill, round the northern edge of which the path winds to the little Kwansakru. This is a woman's village [sic], where the wives of chiefs who have mining-rights, accompanied by their slaves, are stationed, to pan gold for their lazy husbands.

Here, again, the two men use coercion as an explanation. However, this story of hardworking female characters was not meant to invoke pity. There was something more sinister about the context in which it was told. Indeed, Burton and Cameron perceived the activities of these women as deeply troubling, as emphasized by the fictional accounts that they inspired:

> In this way may have arisen the vulgar African story of Amazon settlements. Messieurs Zweifel and Moustier were told by a Kissi man that twelve marches behind their country is a large town called Nahalo, occupied only by the weaker sex. A man showing himself in the streets, or met on the road, is at once put to death; however, some of the softer-hearted have kept them prisoners, and the result may easily be divined. All the male issue is killed and only the girls are kept.
>
> (261)

The treatment of strong women as a threat to masculinity and men comes across quite explicitly in the aforementioned statement.

In the context of the nineteenth century, the concept of "industriousness", which some scholars have equated to a "Protestant work ethic", was, in essence, a societal virtue. It put a spotlight solely on the input side of the work, namely the mind-set, effort, and outlook of the individual/society. Therefore, whereas one can see that European observers had no qualms describing African women, like particular ethic groups, as industrious, this type of diligence would, nonetheless, be coupled with largely negative internal attribution. Furthermore, it is particularly relevant to this volume that although these women could be industrious, for European observers, there was a greater hurdle in the way of viewing them as productive because of their embodiment in terms of race and gender. The concept of productivity was made out to be hard and scientific and to require strategic thinking, and African women workers were excluded from this category on two fronts. First, discrimination occurred on the basis of their being African: There was a widespread belief that before work in Africa could create value, outsiders (mostly Europeans) had to have a hand in this processes. Whether it is gold mining or farming, there first needed to be some type of foreign supervision.[13] Second, this prejudice happened because they were women: It has been argued elsewhere that, whereas productivity was often conceptualized as going hand in hand with "masculine" activities, "feminine" activities could only be regarded as productive with some discomfort (e.g., Colin and Schösser 2013). Perceptions of the work ethics of female African workers were influenced by both categories of race and gender. Arguably, therefore, it would be misleading to tell their stories merely along one of those lines.

In spite of some European travelers and administrators excluding African women from the "lazy African" myth, the counter-stereotype did not necessarily portray female African workers with greater humanity. Industriousness was hardly ever mentioned without classification. In addition, those classifications that were tacked on to such descriptions usually served to discriminate against African women on the basis of race or gender. The limited positive aspects of the counter-stereotype in general and the difficulty of assigning productive attributes to industriousness African women in particular show, in no questionable terms, the still marginal space given to women's work in Africa.

Conclusion

Looked at from a global perspective, histories of productivity reveal critical rifts in the way that economic activity within and without Europe was perceived, shaped, and valued. Yet, the value-based category of "work ethics", as examined in this chapter, stands apart from a productivity that aimed to be measured and treated in scientific terms, as seen in other "regimes of productivity" examined in the book. It is precisely the subjectivity of the matter at hand that allowed for observations to bend in order to meet prejudices of race, gender, and class. Even in cases where the evidence spoke to the contrary, it appears to have been an onerous task for European travelers and administrators to overlook the "lazy African" stereotype. Nevertheless, there were those who presented different accounts of economic life in Africa—accounts that either fully or partially debunked the stereotype.

Using source material produced by male European explorers and officials in Africa, this chapter has attempted to further a dialogue on European perceptions of African work ethics by giving thought to the intention behind counter-stereotypes, in particular, those that highlighted the industriousness of African women. Rather than resting at the point of realizing that some Europeans created a counter-stereotype to the myth of the "lazy African", it brings the tangible impact of the stereotype, the long-standing dehumanizing effect, into the discussion. Most critically, it asks to what extent the counter-stereotype helped to rehabilitate the sullied image of the African worker. Arguably, a successful counter-stereotype would forgo the tactic of using internal attributions, including race and gender, to explain how laborers worked in Africa. Against this background, the nuances are most interesting. An exploration of counter-stereotypes involving the economic activity of African women has served to show that they were not necessarily equal in purpose. If there was no straightforward manner in which Europeans understood or employed the stereotype, just as much variety can be observed among those who constructed the counter-stereotype. Yet, it is the effect of these words that matters most in the end. Indeed, it is within this realm that we can observe that the counter-stereotype was not always conceived as a way to combat preexisting images of African workers. So long as work ethics in Africa were regarded as a function of race and gender, it remained almost impossible for many Europeans to view African laborers as being capable of productivity where colonial supervision and control was absent.

Notes

1 National Archives of the United Kingdom, Kew: CO96/293, Hull to Colonial Secretary, Cape Coast, May 14, 1897, enclosed in GCC 212 of May 27, 1897 (quoted in Austin 2012).
2 In the future, it would be worth undertaking a separate study to explore the perspectives of white European women on the work ethics of African women around this same period.
3 The term "worker" is not being used in a classical Marxist sense here.
4 For example, in her study on the history of au pairs in the UK, Rosie Cox (2015, 261) notes of the colonial period, "racial difference was mapped on to class and gender difference in the homes of the British middle and upper classes".
5 At the Muhlenberg Mission near the Saint Paul River, Reverend Day had opened up a coeducational school where he taught trade classes such as blacksmithing and carpentry. He also established a large coffee plantation in its vicinity.
6 Adam Smith (1776) especially strengthened the abolitionist argument, as he contended that slave labor was more expensive and less productive than free wage labor.
7 The full name of this British club was the Association for Promoting the Discovery of the Interior Parts of Africa.
8 For an original voice, see Esther Boserup (2007).
9 This word has been changed from its original "Fanti" in the text, an older spelling.
10 My thesis, among other things, tells the story of female mineworkers in the Gold Coast's (Ghana) mechanized gold mines in the late nineteenth and early twentieth century. They were strictly hired as casual laborers in surface mining, panning, and washing ore, and they returned to their homes at the end of a day's work.
11 Spelled "Kroomen" in original text.
12 Spelled "Fanti" in original text.
13 It is possible that a perceived lack of political organization (or centralization) in many parts of Africa further impacted how European observers assessed work and the production of economic productivity on the continent.

Bibliography

Alatas, Syed Hussein. 1977. *The Myth of the Lazy Native: A Study of the Image of the Malays, Filipinos and Javanese from the 16th to the 20th Century and Its Function in the Ideology of Colonial Capitalism.* Abingdon: Frank Cass.

Alcoff, Linda Martin. 2006. *Visible Identities: Race, Gender and the Self.* Oxford: Oxford University Press.
Alldridge, Thomas Joshua. 1901. *The Sherbro and its Hinterland.* London: Seeley.
Allen, William. 1841. *A Narrative of the Expedition Sent by Her Majesty's Government to the River Niger in 1841.* Vol. 2. London: Johnson Reprint Corporation.
Atkins, Keletso E. 1993. *The Moon Is Dead! Give Us Our Money!: The Cultural Origins of an African Work Ethic, Natal, South Africa 1843–1900.* London: Heinemann.
Austin, Gareth. 2012. "Vent for Surplus or Productivity Breakthrough? The Ghanaian Cocoa Take-Off, c. 1890–1936." *African Economic History Working Paper Series* 8: 18–9.
——— and Stephen Broadberry. 2014. "Introduction: The Renaissance of African Economic History." *The Economic History Review* 67: 893–906.
Blackburn, Robin. 1997. *The Making of New World Slavery: From the Baroque to the Modern 1492–1800.* London: Verso.
Boserup, Esther. 2007. *Woman's Role in Economic Development.* London: Earthscan.
Boyle, Frederick. 1847. *Through Fanteeland to Coomassie: A Diary of the Ashantee Expedition.* London: Chapman and Hall.
Bridges, Roy. 2002. "Exploration and Travel Outside Europe (1720–1914)." In *The Cambridge Companion to Travel Writing*, edited by Peter Hulme and Tim Youngs, 53–69. Cambridge: Cambridge University Press.
Burton, Richard F., and Verney L. Cameron. 1883. *To the Gold Coast for Gold: A Personal Narrative.* Vol. 2. London: Chatto and Windus.
Büttikofer, Johann. 1890. *Reisebilder aus Liberia: Resultate Geographischer, Naturwissenschaftlicher und Ethnographischer Untersuchungen während der Jahre 1879–1882 und 1886–1887.* Vol. 1. Leiden: Lewin Brill.
Colin, Nicole, and Franziska Schößler. 2013. "Einleitung." In *Das nennen Sie Arbeit? Der Produktivitätsdiskurs und seine Ausschlüsse*, edited by Nicole Colin and Franziska Schößler, 7–28. Heidelberg: Synchron.
Cooper, Frederick. 1987. *On the African Waterfront: Urban Disorder and the Transformation of Work in Colonial Mombasa.* New Haven: Yale University Press.
———. 1996. *Decolonization and African Society: The Labor Question in French and British Africa.* Cambridge: Cambridge University Press.
Cox, Rosie. 2015. "From Our Own Backyard? Understanding UK Au Pair Policy as Colonial Legacy and Neocolonial Dream." In *Colonization and Domestic Service: Historical and Contemporary Perspectives*, edited by Victoria K. Haskins and Claire Lowrie, 256–72. New York: Routledge.
Dumett, Raymond. 1998. *El Dorado in West Africa: The Gold-Mining Frontier, African Labor, and Colonial Capitalism in the Gold Coast, 1875–1900.* Athens: Ohio University Press.
Equiano, Olaudah. 1789. *The Interesting Narrative of the Life of Olaudah Equiano.* London: Self-Published.
Grosz, Elizabeth. 1994. *Volatile Bodies: Toward a Corporeal Feminism.* Bloomington: Indiana University Press.
Hill, Polly. 1997. *The Migrant Cocoa-Farmers of Southern Ghana: A Study in Rural Capitalism.* Oxford: James Currey.
Jeater, Diane. 2006. "The British Empire and African Women in the Twentieth Century." In *Black Experience and the Empire*, edited by Philip D. Morgan and Sean Hawkins, 228–56. Oxford: Oxford University Press.

Mark-Thiesen, Cassandra. 2014. “West African Labour and the Development of Mechanized Mining in Southwest Ghana, c. 1870s to 1910.” PhD diss., Oxford University.

Meredith, Henry. 1812. *An Account of the Gold Coast of Africa: With a Brief History of the African Company*. London: Longman.

Moyd, Michelle R. 2014. *Violent Intermediaries: African Soldiers, Conquest, and Everyday Colonialism in German East Africa*. Athens: Ohio University Press.

Rönnbäck, Klas. 2014a. “‘The Men Seldom Suffer a Woman to Sit Down’: The Historical Development of the Stereotype of the ‘Lazy African.’” *African Studies Journal* 73: 211–27.

———. 2014b. “The Idle and the Industrious—European Ideas About the African Work Ethic in Pre-colonial West Africa.” *History in Africa* 41: 117–45.

Smith, Adam. 1776. *The Wealth of Nations*. London: W. Strahan and T. Cadell in the Strand.

Yeats, John M. 2011. “In Praise of Industry: Early Nineteenth-Century Concepts of Work.” *Journal of Markets & Morality* 14: 147–57.

4 Saving the Supply and Making People Work

Sustainability, Labor, and Control of Production in the Rubber Trade of Southeast Cameroon, 1899–1903

Tristan Oestermann

In late 1898, the German Kolonial-Abteilung (Colonial Office) granted a giant land concession in the southeast of the colony of Cameroon to a German-Belgian joint venture named Gesellschaft Süd-Kamerun (South Cameroon Company, GSK; see Ballhaus 1968; Oestermann and Geschiere, forthcoming).[1] This area encompassed one-fifth of the entire colony and was covered with little known and thinly populated forests. The company's goal was to exploit the land's rubber riches. As there was no permanent overland connection to government stations at the coast, the only access to this remote region of Cameroon was via the Congo and Sangha rivers. From here, the GSK's agents could enter the Ngoko/Dja, the only river at least partly navigable in the concession area. The company started its trading business at the river's banks in April 1899. From its headquarters in the riverside village of Moloundou, a network of factories[2] and trading posts spread over the region where its agents intended to buy rubber from the small, multiethnic population. Despite these efforts, trade developed at a slow pace: It proved very difficult for the GSK to make people produce rubber and work in its caravans. In October 1900, one of the company's experts traveling on the Dja noted about the local Missanga people, "none of the men has wanted to make rubber". They "don't make rubber, don't want to work as carriers, and hardly go out to search ivory".[3]

In the following years, the GSK was able to significantly increase its trade in rubber as people became more and more willing to gather the raw material, whereas the lack of porters remained a pressing problem. The company could even pay a small dividend in 1903. However, this success was only possible because of significant support from the colonial administration that helped to mobilize local African labor for the GSK's profit. As stated in the instructions of the first colonial officials of the region, the main function of the state in the concessionary area was to ensure its exploitation by the GSK,[4] who had to pay a certain percentage of its profit to the state (Ballhaus 1968, 106). The GSK's state support included the use of coercion and military force. Therefore, as in other regions affected by the global rubber boom in the early twentieth century (see e.g., Coquery-Vidrovitch 1972 for Africa; Weinstein 1983 for Amazonia; Hochschild 1998), violence and forced labor played an important role in the making of the rubber export economy of southeastern Cameroon. Yet why exactly did local colonial rulers opt for violence, coercion, and control? This chapter argues that force played such an important role because, in the first period of the GSK's existence (1899–1903), the colonial authorities and the company tried to establish a colonial regime that could control the entire production process.

Presumably, because colonial authorities and the GSK saw control as the decisive precondition for further steps, during this quest for control, they did not use the term productivity in statements about the production of rubber and the work of the colonized people. Rather, they employed notions of proto-sustainability and drew heavily on racist colonial discourse to make sense of African economic choices that did not match their expectations. Notions contrasting Africans as "irrational" and "idle" and Europeans as "rational" and "industrious" (see Mark-Thiesen, this volume; Eckert 1999; Cooper 2006) were used to legitimize violence. This chapter will examine this phenomenon by analyzing the handling of two of the GSK's major concerns: first, the introduction of a European-style model of "rational" rubber production in contrast to allegedly typical "African" overexploitation and second, the problem of labor shortage, which was particularly pressing in terms of carriers.

Colonial officials and traders claimed that African economic activity could not be productive without European control or coercion. So before colonial experts started to elaborate a notion of productivity, colonial control had to be achieved. Nevertheless, the problems the GSK still faced in 1903 and in the following years, especially concerning the lack of porters, indicate that the colonized continued to make their own economic choices, despite all violent attempts to control them. It is not easy to reconstruct these choices. Hardly any African voices from southeastern Cameroon survived in the primary sources. Because oral history has barely been conducted (see Geschiere 1982; Giles-Vernick 2002; Rupp 2011; Zouya Mimbang 2013 for oral history), at this point, the major sources remain colonial reports that have to be interpreted carefully in order to allow for the reconstruction of African views about the rubber boom. As will be made plausible, in the GSK-concessionary zone, people actually grasped the new opportunities the rubber trade was offering them. Yet they mostly refused to work as carriers for the company because of the better benefits of their labor as rubber collectors.

Raubbau or "Rational" Production: Controlling Rubber Production in Cameroon

When the GSK was expanding its trading business in the concessionary area, both the firm and the government insisted that "rational" methods be used to produce the rubber that the company bought in its factories, in contrast to "African" Raubbau (overexploitation—literally meaning to rob resources). Because "rational" production promised to provide a better product and a sustainable supply of rubber, as well as to serve as an argument against the admission of competitors into the region, the authorities and the company introduced practices that should have prevented the dieback of the rubber trees, although often without success. Where colonial administrators came across the use of destructive methods in the far-flung forests, they applied force to make people gather rubber according to their model. Thus, the very notion of "rational rubber production" was based on racist ideas about Africans and the necessity for coercion.

In order to understand the importance of the concept of "rational" exploitation, it is useful to examine the history of West African rubber production. Before and during the rubber boom, most of the world's rubber was gathered in Brazil by tapping Hevea brasiliensis trees in the Amazonian rain forests. However, because of changes in the industrial use of rubber, global demand exceeded what Latin America could supply (Tully 2011). Increased prices led European merchants to look for alternative sources of rubber on the western coasts of Africa—a demand to which Africans readily responded. As Cuthbert Christy (1911) and Robert W. Harms (1975, 73–75) presented in their overviews, Angola and Gabon began to export rubber in the 1870s, followed by Sierra Leone, French Guinea, and the Gold Coast in the 1880s, and southern Nigeria and Cameroon in the 1890s. In these regions, the rubber trade was controlled by African entrepreneurs mediating between the producers in the forests and European traders in their factories on the coast (see Dumett 1971; Wirz 1972; Omosini 1979; Osborne 2004; Vos 2009; Fenske 2010).

In all these regions, high prices and rapidly expanding demand led to a boom-and-bust-style economy, best illustrated by the example of southern Nigeria: In 1894, the colony exported only three tons of rubber. In the following year, the amount was already 2,263 tons, which led to the degeneration of 75 percent of all rubber trees (Whitford and Anthony 1926, 22; Omosini 1979, 27–9): People felled the Funtumia elastica trees or cut off the rubber-yielding Landolphia vines, thereby killing the plants. Because of this overexploitation, trading frontiers quickly moved inland as soon as the rubber boom set in (Harms 1975, 74–5).

Cameroon's coastal region suffered from a similar pattern. As early as 1889, when the colony's rubber trade was still negligible, Governor Julius von Soden wrote to Chancellor Otto von Bismarck that "given the way the native population produces rubber, the total destruction of all rubber-yielding vines and trees and, therefore, our rubber trade will only be a matter of time".[5] The problem of Raubbau gained particular importance at the Batanga Coast, which increasingly specialized in the export of rubber. Despite the destructive consequences, in the dynamic of the boom, African producers continued to fell trees and cut vines. These practices quickly produced large quantities of rubber and, thus, large profits. When local supplies were destroyed, the rubber frontier moved farther into the interior of the colony. In the mid-1890s, factories were opened in the immediate hinterland of the coast, and in 1899, the furthermost factories of the coastal firms were already established in Yaoundé; they entered the far-off concession of the GSK in 1903. Until its abrupt end in 1913, when wild rubber became irrelevant because of the advent of Southeast Asian plantation rubber, the rubber trade of southern Cameroon could rightly be termed Raubwirtschaft: a robber economy destroying its own future.

In 1898, the year that saw the founding of the GSK, Raubbau showed its consequences: Rubber exports from all African colonies (except the Congo Free State where forced labor was used to open up new districts; see Harms 1975) were declining, while the rubber industries in Europe and the USA were still booming. Interestingly, German scholars and colonial administrators, fearing a scarcity of the raw material, did not start to think about measures to raise productivity to obtain more rubber from the forests. Instead, they began to discuss methods to secure Cameroon's production in rubber, that is, by extending European control of the production process through the supplanting of overexploitation by sustainable production (Warburg 1897; 1900, 251; Tropenpflanzer 1898; Preuss 1899; Treue 1955).

Concepts of overexploitation and sustainable production had a long history in German science. It began in the eighteenth century, when scientific forestry emerged in order to stop the depletion of woods. In the middle of the nineteenth century, Justus Liebig popularized the highly moral metaphor of Raubbau that was soon used by other scientists to condemn the exhaustion of natural resources (Raumolin 1984). Generally, rational use of the soil and raw materials was presented as a remedy. In the colonies, science and reason were joined by an "environmental chauvinism" (Rollins 1999, 198) that claimed European superiority in guarding the land's riches. Accordingly, only Europeans could save these resources through the application of "rational" scientific methods.[6] Raubbau and the ability to abandon this trajectory were, thus, directly connected to racial and cultural hierarchies. In a Darwinist way, the geographer Ernst Friedrich (1904, 70–5) saw Raubwirtschaft and its repercussions as a part of the struggle for survival and a necessary step on the way to rational production, progress, and civilization. Writing about Africans overexploiting rubber trees, he characteristically did not even consider whether Africans themselves were able to leave behind Raubbau, invent sustainable systems, and create progress, but reasoned that Europeans—because of their higher degree of civilization—were compelled to create rubber plantations in order to secure a sustainable supply and, thereby, would contribute to progress.[7]

Against this background of "reason" and racism, overcoming Raubbau in Cameroon by gaining control over the production process was presented as a challenge for Europeans. Famous botanists such as Otto Warburg and the director of Cameroon's botanical garden in Victoria, Paul Preuss, suggested the establishment of European-owned rubber plantations in order to obtain a sustainable rubber supply (Preuss 1899, 17; Warburg 1900). It was clear to them, however, that this would only help in the long run. Others argued for more short-term actions to save the existing trade, aiming not at the control of the means of production but of the colonized population's actions. Yet the idea to systematically punish Raubbau was quickly overturned. As colonial rule in Cameroon at the time was still limited to the coasts and some lonely inland stations, an official wrote in a memorandum: "The dense primeval forest eludes every form of control of the methods of rubber production".[8] Deputy Governor Theodor Seitz, therefore, proposed that colonial firms take responsibility for Raubbau and teach the African population "rational" methods of rubber production, such as tapping the trees and vines, instead of chopping them off.[9]

Yet the prospect of teaching "rational" methods to Africans was contested. Preuss, who was considered an influential authority, preferred to establish plantations; he mixed his personal experience of African Raubbau at Barombi station and his knowledge of the depletion of rubber on almost every African rubber frontier with racist assumptions about limited African intellectual potentials. "The" African, he declared, "indeed sees that he kills the hen which lays golden eggs when he chops off the rubber tree, but he can hardly fight against his nature. He is a child of the moment and a safe huge profit today is more important to him than the prospect of a smaller profit in the future, may it even repeat itself frequently" (Preuss 1899, 16).

In this view, overexploitation was not simply a matter of method but a matter of race. Africans, who were reduced to their bodies and were denied any higher intellectual capacities, were, thus, depicted to be able to use "rational" methods of exploitation only when being instructed and controlled by Europeans—although most European merchants did not know any better than their African producers.[10] In the end, because the state lacked the funding and the merchants lacked interest, people were not taught any methods of "rational" exploitation along the Batanga Coast nor in its neighboring hinterland (Wirz 1972, 108–10).

When the GSK started its trading operations in southeastern Cameroon in April 1899, conditions were different from those at the coast: The Belgian predecessors of the GSK, who were the first European merchants entering this remote region, did not trade in rubber, mainly because of the high transportation costs before the opening of the Congo Railway in 1898. They confined themselves to buying ivory, which had a higher value per item.[11] Whereas colonial administrators even claimed that making rubber was unknown to the population, oral history interviews suggest that rubber might have been used in small quantities for local purposes, but it seems implausible that it was used as a commodity.[12] For this reason, the destructive methods for maximum short-term profit developed at the coast were not implemented. These conditions retarded growth in the rubber trade of the concession area: It was only by 1900 that the GSK began to export the first few tons of rubber.[13] Nevertheless, colonial authorities and the GSK took advantage of the local lack of destructive methods and established a model of "rational" exploitation informed by colonial botanists from the metropolis in order to ensure the sustainable production of rubber.

Initially, neither the colonial administration nor the GSK intended to introduce this model. At first, the rubber-rich southeast was most likely only opened for commerce so as to boost rubber exports after the decline of coastal production. While planning the business model of the GSK, its later director, Julius Scharlach, disclosed to one of his partners that there was actually no alternative to "Raubbau".[14] Furthermore, both government and company proved rather helpless in the face of people's reluctance or inability to produce large quantities of rubber.[15] Yet they were lucky: Early in 1899, the colonial lobby organization Kolonial-Wirtschaftliches Komitee (Colonial Economic Committee) decided to send the young botanist Rudolf Schlechter to West Africa in order to explore the state of the rubber industry, the methods of production, and the best species for rubber plantations in the German colonies. Over the course of his Westafrikanische Kautschuk-Expedition (West African Rubber Expedition), Schlechter (1900b) visited southern Nigeria and the Congo Free State. From there, he traveled into the GSK's concession in September 1899. Even though the main aim of the expedition was to initiate plantation production of rubber, Schlechter's work in the southeast was of the greatest importance for the development of the GSK's regime of production. Drawing on his theoretical knowledge, as well as on his experience and experiments, Schlechter introduced what he claimed to be "rational" rubber tapping in the concessionary zone: The bark of Funtumia elastica trees was cut in a fish bone pattern. This method was utilized to produce a profitable yield in rubber while keeping the tree alive for future use. Schlechter (1900b, 100) showed his methods to the African workers of the administration and the GSK and also to the equally inexperienced European members of the staff.[16]

Both the administration and the company embraced the new techniques and made them a central part of their regime of production. In the following years, the firm's agents and colonial officers actively encouraged rubber production using Schlechter's method in all parts of the concession. On their joint expeditions through the unknown territories, they gave "a practical demonstration of rubber production to the completely ignorant inhabitants of the country in every bivouac".[17] When expeditions came across Raubbau in the concession area, colonial actors resorted to coercion in order to enforce "rational" exploitation. Indeed, Schlechter himself made clear that coercive methods were to be used to save the country's capability to produce rubber from the assumed unreasonableness of the African population:

> One of the first and most important tasks of the district officer and the merchants in the Ngoko district will be to show the natives the value of rubber and its production, and, if possible, in such a way that

> overexploitation does not become a habit in this region, too. Indeed, I fear that it will not be possible to keep overexploitation at bay even by using strict measures. But despite this, it would be desirable to strictly punish the felling of Kickxia trees [Funtumia elastica] when it is verifiable.
>
> (Schlechter 1900b, 107)

Just as Schlechter proposed, colonial authorities severely prosecuted the destruction of rubber trees, when they themselves or the GSK's agents came across it at all, given the vastness of territory. Therefore, punishment and coercion were rather punctual and very harsh so as to set an example: Entire villages, rather than individual "perpetrators", were held in account for Raubbau. District Officer Captain Ludwig von Stein, for instance, collectively convicted villages of the Konabembe people and enforced punitive payments for chopping off rubber trees. Usually, those payments included quantities of rubber and ivory and the appointment of forced laborers.[18]

Until 1903, the "rational" method of rubber tapping was introduced into the majority of the concession. Yet, in the same year, traders from the Batanga Coast first entered the area on an overland route and brought their destructive methods with them (see Ballhaus 1968, 146). In the ensuing fierce and violent competition between the GSK and the Batanga traders, the concept of "rational" tapping became a keystone in the concessionaires' strategy to defend their former de facto trading monopoly against its rivals: The company's main argument of defense was its insistence on "rational" rubber production in contrast to the Raubbau other merchants allegedly were inciting and practicing. Again, the GSK mobilized the racist implications of the concept of "rational" exploitation. It insisted on the need for strict control of the production process by the concessionary company to prevent people from overexploiting the trees.[19]

Thus, with the help of the concept of "rational" rubber production, the GSK tried to install an economic system based on control and coercion so as to sustainably produce rubber in its concession. The racist bias inherent in this model legitimized violence in order to enforce control over the production process. Because of the lack of funds and staff, the immensity of the territory, and its geographical inaccessibility, though, full control was not achieved. Nonetheless, after a shaky start, rubber production increased, not because of but despite violence and coercion, as the following section will demonstrate.

Wealth in Rubber or Wealth in People: The GSK and the Arbeiterfrage

Making people gather rubber "rationally" was not the only problem the GSK was facing. Like other economic enterprises during German rule in Cameroon—trade, plantations, or railway construction—the GSK also suffered from a lack of labor (Rudin 1938; Rüger 1960; Winkler 1960; Wirz 1972; Mandeng 1973). This Arbeiterfrage (laborer question) posed a serious threat to the concession's capability to produce a profit. The main problem was the recruiting of porters to transport rubber and merchandise within the network of factories and to navigable rivers. One reason for the acuteness of the Arbeiterfrage for the GSK was the local population's own economy of labor: For the greater benefit, people tapped rubber and sold it in the company's factories but were reluctant to work as porters for the GSK. Nonetheless, instead of paying wages high enough to lure people to porterage, the firm and colonial authorities used the racist stereotype of African "laziness" to denounce economic choices and to justify forced labor.

The low population density in those parts of the colony that were economically relevant for the colonialists (Mount Cameroon, the coast, and the south) was the main reason German Cameroon suffered from a lack of workers. Density was lowest in the rain forests of the GSK concession where statistically less than one person lived on a square kilometer (see e.g., Koch 1913, 282; Vansina 1990, 46; Hoffmann 2007, 1: 147). Company agents and administrators had to cope with what they called "dead zones": forests that stretched up to a fourteen-days journey without any village to buy food, giving the region the Pidgin-English nickname "hungry country" (Haase 1915, 90).[20] Consequently, the new rulers—companies and administrators alike—had to learn that wealth in this part of Africa was not founded on the control of land but the control of people (Guyer 1993, 243–5). Although the GSK owned a gigantic territory rich in rubber, it was worthless without workers harvesting and transporting the raw material.

Despite the coercion connected with the "rational" model of tapping, the GSK succeeded in its effort to make people gather rubber largely because of its capacity to hijack an already existing socioeconomic system. As of 1902, the rubber trade began to flourish. People tapped rubber voluntarily, brought tons of it to the factories, and exchanged it for European import goods such as wire, flintlocks, knives, brass spirals, cloth, and other items that had little value in the colonial metropolis. In 1900, European merchants were able to buy one kilogram of rubber for one Mark, later for two Marks in southern Cameroon, but could sell it in Hamburg for up to 7.15 Marks per kilogram.[21] Whereas Marxist historians in the 1960s had identified this as a clear case of exploitation (Rüger 1960; Ballhaus 1968; Kaeselitz 1968), Albert Wirz demonstrated that people's acceptance of these conditions can be explained by the fact that in West and Equatorial Africa, many European trade items were part of a huge number of currencies. Most characteristically, they played a crucial role as special-purpose money in marriage payments and, thus, the quest for wealth in people (Wirz 1972, 103–04; Guyer 1986; 2004, 15). In precolonial times, European import goods were acquired through ivory trade. Access to elephant tusks was, then, restricted to Big Men and chiefs, because only they were part of trading networks, had links to local Baka hunters, or were able to mobilize enough men to kill an elephant. As a result, the ivory trade strengthened the gerontocratic structures of the societies in southeastern Cameroon. All young men relied on the goodwill of elders for the permission to marry and advance socially (see Harms 1981, 41; Laburthe-Tolra 1981; Oestermann and Geschiere, forthcoming).

The advent of the rubber trade radically changed this situation: Tapping rubber and bringing the yield to the GSK's factories made it possible to acquire European goods independently from Big Men and chiefs. Therefore, young people especially grasped the opportunities of the new rubber trade and the possibility to join the race for control of people on their own account (Wirz 1972, 107; Geschiere 2007, 51–3). The attractiveness of the new business is evidenced by the fact that people very quickly invented new techniques to improve rubber tapping: Konabembe-men developed special knives to cut the Funtumia, and the Mkako designed a special device to facilitate climbing the trees.[22] Without noticing it, the GSK was lucky to tap into a social system that included an intrinsic motivation for many people to gather rubber for the company by making it possible for them to pursue a strategy of what Arjun Appadurai (2012, 25–6) has called "diversion".

However, as much rubber as the GSK could amass, in colonial Cameroon, the crucial question remained transport: On the one hand, the country lacked navigable rivers that allowed easy access to the hinterland. The streams flowing to the coast, as well as those in southeastern Cameroon flowing into the Congo, were blocked by rapids, hindering the use of steamers and, often enough, even pirogues for river traffic. On the other hand, the tsetse fly quickly killed all beasts of burden. So, despite its high costs, the only way to move rubber or any other merchandise in the hinterland was the use of manpower. The exploitation of the GSK's concession, therefore, depended on the bodies of the African population that had to carry its goods and products. In 1903, the company continuously needed more than seven hundred laborers; over three hundred of them were porters connecting the far-flung network of factories and posts in large caravans of up to ninety people.[23] Tons of rubber had to be transported by their muscular strength to the company's headquarters in Moloundou, where it was possible to use the river Ngoko to ship the produce via Sangha and Congo to the open sea.

However, the company was hardly capable of recruiting local porters, even though porterage was also a way to earn local currencies.[24] Schlechter (1900a, 114) reported in 1899: "Until now, all attempts to persuade the natives to services—even by comparatively high wages—failed". Regularly, rubber piled up in the far-flung factories. In 1906, the situation was so bad that rubber in the factory of Yokadouma had been stored for two years and had already begun to rot because the GSK did not have enough porters to move it to Moloundou—a twelve-day march away.[25] To get the rubber moving, the company was forced to "import" expensive migrant laborers, especially Vai from Liberia, but also Bangala from the near Congo, right from the start. They formed the core of the rubber caravans but were a huge financial burden (see Jäckel 1909, 276; Koch 1913, 259).

The GSK depended on this imported caravan labor because, unlike other parts of Cameroon, large, long distance caravans were uncommon in the southeast. As people were organized in small independent village communities, and the political environment was marked by instability, trade was conducted from village to village. Thus, relatively small groups of people moved only short distances for trading. Ivory, the main trade item, passed through many hands while it was moved to Yaoundé or into the Congo basin (Wirz 1972; Harms 1981). So when the GSK started trading in its concession, there was no established labor force of porters it could rely on. Furthermore, the societies of southern Cameroon regarded porterage (especially for people other than kin) as slave labor (Geschiere 1982, 143–4; Guyer 1993, 255). In 1900, an expert of the GSK reflected this attitude in his writings about the Bangando:

> These people don't want to work for strangers. They regard it as a humiliating work . . . The first caravan from Tschimbuli na n'xé to Moloundou complained vividly that inhabitants of the other villages had received them very badly and reproached them for doing boy-services for the white people.[26]

Additionally, becoming a carrier was not attractive in southeastern Cameroon because of the poor conditions of porterage. Laborers were absent from home for long periods, sometimes for several months. They had to carry heavy loads of thirty-five kilograms and more along narrow paths through swampy rain forests, often enough sinking deeply into the mud; they were exposed to heavy rainfall and exhaustion.[27] At times, nutrition was scarce, as no food was available in the "dead zones". As one detailed report of a caravan of the GSK indicated, the regime of control was very rigid and based on violence: Even minor offenses were punished by beating or enchainment.[28]

Given the low density of population in the forest area, the local understanding of porterage as low-status labor, and the hardship of life as carriers, the price to engage people in caravan work was very high. In 1903, the GSK paid each "native porter" twenty Francs (approximately 16.20 Marks) in goods for traveling from Yokadouma to Moloundou and back, which took about a month—a quite luxurious wage, given the fact that workers on cocoa plantations on the coast earned only six to ten Marks in goods a month (Rüger 1960, 223–4). Yet still, people were reluctant to work as carriers, and in 1905, Bangando porters complained about low wages.[29] The GSK-agent C.W.H. Koch and the official Leo Treichel presented a simple reason for this. Both accounts were written several years later, at the height of the rubber boom, but their findings supposedly apply to earlier times as well: By tapping rubber, people could earn as much or even more without suffering the hardship of porterage.[30] Even though the terms and notions the colonized used to justify their choices are unknown, one can assume that they chose collecting rubber over other economic activities to reach a higher profit for their labor and, thereby, advantages in their efforts to marry and control people themselves.

Yet colonial authorities and the concessionary company were not willing to interpret their actions as socially and economically rational. Rather, in the eyes of colonialists, this behavior confirmed the stereotype of the "lazy native". To cite only one, C.W.H. Koch, who gave an appropriate account of why people chose rubber tapping over porterage, told his readers in the same article that "laziness" was one of "the native's" most important characteristics. About the Nzime-people, Koch declared: "Work is what they dread most" (Koch 1913, 275, 281).

As people refused porterage and the costs for "imported" West African migrant workers were too high, the company began to draw on the racist concept of Erziehung zur Arbeit (education for work—see e.g., Conrad 2010, chap. 2). Yet the missions that used to implement the concept in other parts of the German colonial empire were not active in the southeast. Thus, for the GSK, Erziehung zur Arbeit meant nothing but forced recruitment of porters to show people the "benefit" of work and to make them get "used to it".[31] Given the colonial administration's interest in the development of the GSK and the concession, it was fairly willing to lend a hand to the company. Very soon after Governor Jesko von Puttkamer legalized the recruitment of workers by the German colonial army in 1899, so as to solve the notorious Arbeiterfrage, the authorities began to supply forced laborers to the GSK (see e.g., Stein 1902, 442; Rüger 1960, 195–6). Additionally, military expeditions to extort Strafarbeiter (punitive laborers) from villages that did not submit to the new German rulers, served to recruit porters to the GSK (Rüger 1960, 201; Wirz 1972, 139). As an example, in 1901, von Stein led an expedition against the chiefdom of Bertoua, which was forced to provide one hundred Strafarbeiter. As the conscripted managed to escape, von Stein extorted another hundred men from Bertoua.[32] The consequence of this policy was that until 1903, the GSK became more and more dependent on forced laborers as part of its regime of production, a factor that remained of considerable relevance until the outbreak of the Great War (Petersen 1939).

Conclusion

The GSK and the colonial administration tried to establish a regime of production on the racist assumption that African economic activity could be neither sustainable nor productive without European coercion and control. First, based on a discourse of proto-sustainability and Darwinism, they argued that Africans would only harvest rubber "rationally" when the production process was sufficiently controlled by Europeans. Only force was thought to keep up "rational" exploitation of rubber, even though people quite readily learned and adopted the new techniques. This was the case not least because the rubber boom offered Africans opportunities for profit within their socioeconomic system: access to local special-purpose money and, as a result, marriage and social advancement. For this reason, they were keen to adopt the model of "rational" rubber production forcefully introduced by the Europeans.

Second, the colonial actors were certain about the inherent idleness of the African worker. Therefore, they interpreted the reluctance of people to work as porters as a racial character trait. Even though people were working hard tapping rubber and bringing it to the GSK, the colonialists believed that people were lazy and could only be mobilized by force—at least for porterage. Yet, considering the evidence, it can be assumed that the colonized were making well-founded economic choices, because collecting rubber was culturally and materially much more valued than porterage. Thus, people could be more successful within their social and economic system when they confined themselves to rubber production. Nonetheless, the colonial power used the military to coerce people to work as carriers, instead of paying higher wages.

Hence, the situation ensuing in southeastern Cameroon can be interpreted partly as the consequence of the violent interaction of two economic regimes: on the one hand, the quest of the colonized population for the most profitable work, seizing the opportunities of rubber tapping but refusing porterage, and, on the other hand, the colonial regime of production, adhered to by the GSK and the colonial administration, resting on the belief that Africans lacked rational economic ideas and were too "lazy" so that African labor needed to be controlled. Given these conflicting views, the GSK's regime of production rested on the support of the colonial authorities. Only with the military's help regarding the Arbeiterfrage was the GSK able to render its concession successful and to keep up its trading business.

However, even though the GSK appreciated that colonial authorities coerced people to work, thereby keeping wages relatively low and sparing the company the employment of expensive African migrant workers, it is important to note that this was not a measure to raise productivity. Rather, it was a means to ensure that rubber could be traded at all and, therefore, part of the quest of colonial actors to attain control over the economic life of the colonized. It might well have been the case that the GSK began considering measures to raise productivity after its concession was reduced in 1905 to one of the "dead zones". There, the GSK radically changed its mode of operation: Instead of trading rubber, it sent gangs of African workers (many of them forced laborers) into the dead zones to employ "rational" production of rubber (Petersen 1939). It would be very interesting to ask whether, in this regime of production, notions of proto-sustainability and racist discourses allowing for the violent control of African economic behavior gave way to questions of productivity. Even though the necessity of control for regimes of productivity was contested in the twentieth century (see Alexa

Geisthövel and Lukas Held, this volume), the question of how concepts of productivity were linked to racist discourses, but also to notions such as sustainability, needs further analysis with respect to colonial contexts and beyond.

Notes

1 I would like to thank Peter Geschiere, Janina Rühl, Christine Strotmann, Philip Gater-Smith, and not least the editors for comments on the present paper, parts of which are based on my unpublished master's thesis, supervised by Andreas Eckert.
2 Factories were the name given to trading compounds on the West African coast in the nineteenth century. They normally consisted of several buildings to lodge the commercial agent and his African workers, a kitchen, stables, stores, and depots for the purchase of produce and the sale of merchandise. A trading post, however, might only be a small African-style hut in a village. See i.a. Wirz 1972, 115–6.
3 Bundesarchiv Berlin (BArch) R 1001/3443: Williame to Delcommune, Molundu, 27/11/1900, 83–6. All translations are mine.
4 BArch R 1001/4374: Puttkamer to Plehn, Kamerun, 15.10.1898, 137–9.
5 BArch R 1001/8113: Soden to Bismarck, São Thomé, 20/07/1889, 10; see Wirz 1972, 108–10.
6 Whereas William H. Rollins (1999, 198) uses the notion of "environmental chauvinism" for the tendency to essentialize environmental awareness as exclusively German, this chapter employs it in a broader sense of alleged European superiority.
7 For an analysis of Friedrich's article, see Jussi Raumolin (1984, 801–3).
8 BArch R 1001/8113: Memorandum of von Oberndorff, Kamerun, 14/01/1900, 291. A ban was decreed only in August 1906 (see Wirz 1972, 109).
9 BArch R 1001/8113: Seitz to Colonial Office (CO), Kamerun, 16/08/1898, 204.
10 See, for instance, BArch R 1001/8114: Stein to CO, Berlin, 15/02/1904, 74.
11 Historisches Archiv der Deutschen Bank (HADB) K1/395: Memorandum of Albert Thys, not dated (1898).
12 BArch R 1001/8114: Stein to CO, Berlin, 15/02/1904, 75. Lucie Zouya Mimbang (2013, 139) mentions rubber tapping but does not give examples for precolonial use (see also Wirz 1972, 108).
13 BArch R 1001/8114: Stein to CO, Berlin, 15/02/1904, 75; BArch R 1001/3443: 9th session of the GSK-board, Hamburg, 05/11/1900, 161.
14 HADB K1/395 Scharlach to Schinckel, Hamburg, October 1898.
15 See Plehn to CO, Ngoko-Station, 24/07/1899, 267.
16 BArch R 1001/8114: Stein to CO, Berlin, 15/02/1904, 74–5. Only a few years later, Schlechter's methods were denounced to be even more destructive if executed incorrectly. See BArch R 1001/8114: Schmidt to Puttkamer, Buea, 26/02/1905, 214–17; Schulte im Hofe (1906, 147–8).
17 BArch R 1001/8114: Stein to CO, Berlin, 15/02/1904, 75.
18 BArch R 1001/8114: Stein to CO, Berlin, 15/02/1904, 77.
19 See e.g. BArch R 1001/3445: Report Scharlach, Hamburg, 01/03/1903, 13–6.
20 See Ballhaus (1968, 147), BArch R 1001/8114: Stein to CO, Berlin, 15/02/1904, 76. The forests were not uninhabited: They were the home of nomadic Baka groups, contemporarily called "pygmies". Additionally, the land formed a reserve for the shifting agriculture of the populations (Vansina 1990, 43–6; Wirz 1972, 93–4).

21 For rubber prices in Hamburg, see BArch R 1001/8114: Weber to Scharlach, Hamburg, 15/04/1913, 15. In Cameroon, prices were calculated in European currencies but paid in goods, leaving room for price manipulations (see Wirz 1972, 123–4; Kaeselitz 1968, 2:37).

22 BArch R 1001/8114: Stein to CO, Berlin, 15/02/1904, 81; Erich R. Petersen (1939, 71).

23 Archiv der Freiherrn von Stetten (AFS) bundle 415: Minute of GSK board meeting, 05/02/1903.

24 Whereas Stephen J. Rockel (2006, 85) claims the attractiveness of porterage for the Nyamwezi-porters of East Africa despite the hardship to be endured, this is certainly not valid for the people of southeastern Cameroon during the period analyzed here. The people recruited for porterage locally showed the characteristics of what Rockel terms "amateurs"—in contrast to the Liberian and Congolese workers employed by the GSK who can be labeled "professionals".

25 BArch R 1001/3443: GSK: report on the first financial year, 05/11/1900, 39–40; BArch R 1001/3443: Williame: Report on the Bumba river, Ngoko, 01/04/1901, 131; AFS bundle 415: GSK to Stetten-Buchenbach, Hamburg, 19/03/1906.

26 BArch R 1001/3443: Williame: Report on the Bumba river, Ngoko, 01/04/1901, 131–2.

27 In 1908, colonial authorities decreed that a porter's load should not exceed thirty kilograms of commodities plus five kilograms of foodstuff. It can be assumed that loads were often much heavier (see Hausen 1970, 269). Regarding porter recruitment, see Adolf Rüger (1960), Patrice Mandeng (1973) and Albert Wirz (1972, 136–7).

28 AFS bundle 344: Diary of Richard von Stetten, 1906.

29 BArch R 1001/3447: GSK to CO, Hamburg, 22/05/1905, 147–51.

30 Koch (1913, 282); BArch R 1001/8118: Treichel to Gouvernement, Djah-Posten, 04/01/1911, 171.

31 BArch R 1001/3448: Münster-Schultz/Schwonberg to CO, Hamburg, 25/10/1905, 120; BArch R 1001/8114: Weber to Scharlach, Hamburg, 15/04/1903, 14.

32 BArch R 1001/3444: Schlippenbach to GSK, Beri, 17/08/1902, 125–33.

Bibliography

Appadurai, Arjun. (1986) 2012. "Introduction: Commodities and the Politics of Value." In *The Social Life of Things: Commodities in Cultural Perspective*, 10th ed., edited by Arjun Appadurai, 3–63. Cambridge: Cambridge University Press.

Ballhaus, Jolanda. 1968. "Die Landkonzessionsgesellschaften." In *Kamerun unter deutscher Kolonialherrschaft: Studien, vol. 2*, edited by Helmuth Stoecker, 99–190. Berlin: Rütten & Loening.

Christy, Cuthbert. 1911. *The African Rubber Industry and Funtumia Elastica ("Kickxia")*. London: J. Bale, Sons & Danielson.

Conrad, Sebastian. (2006) 2010. *Globalisation and the Nation in Imperial Germany*. Translated by Sorcha O'Hagan. Cambridge: Cambridge University Press.

Cooper, Frederick. 2006. "African Labor History." In *Global Labor History: A State of the Art*, edited by Jan Lucassen, 91–116. Bern: Lang.

Coquery-Vidrovitch, Catherine. 1972. *Le Congo au temps des grandes compagnies concessionnaires, 1898–1930*. Paris and Le Haye: Mouton.

Dumett, Raymond. 1971. "The Rubber Trade of the Gold Coast and Asante in the Nineteenth Century: African Innovation and Market Responsiveness." *Journal of African History* 12: 79–101.

Eckert, Andreas. 1999. "Arbeitergeschichte und Geschichte der Arbeit in Afrika." *Archiv für Sozialgeschichte* 39: 502–30.

Fenske, James. 2010. "'Rubber Will Not Keep in This Country': Failed Development in Benin, 1897–1921." *MPRA Paper* 23415.

Friedrich, Ernst. 1904. "Wesen und Verbreitung der 'Raubwirtschaft.'" *Dr. A. Petermanns Mitteilungen aus Justus Perthes' Geographischer Anstalt* 50: 68–79, 92–5.

Geschiere, Peter. 1982. *Village Communities and the State: Changing Relations Among the Maka of South-Eastern Cameroon Since the Colonial Conquest*. London et al.: Keagan Paul International.

———. 2007. "Regional Shifts: Marginal Gains and Ethnic Stereotype." *African Studies Review* 50: 43–56.

Giles-Vernick, Tamara. 2002. *Cutting the Vines of the Past: Environmental Histories of the Central African Rain Forest*. Charlottesville et al.: University Press of Virginia.

Guyer, Jane I. 1986. "Indigenous Currencies and the History of Marriage Payments: A Case Study from Cameroon." *Cahiers d'Études africaines* 26: 577–610.

———. 1993. "Wealth in People and Self-Realization in Equatorial Africa." *Man* 28: 243–65.

———. 2004. *Marginal Gains: Monetary Transactions in Atlantic Africa*. Chicago: University of Chicago Press.

Haase, Lene. 1915. *Durchs unbekannte Kamerun: Beiträge zur deutschen Kulturarbeit in Afrika*. Berlin: Fleischel.

Harms, Robert W. 1975. "The End of Red Rubber: A Reassessment." *Journal of African History* 16: 73–88.

———. 1981. *River of Wealth, River of Sorrow: The Central Zaire Basin in the Era of the Slave and Ivory Trade 1500–1891*. New Haven: Yale University Press.

Hausen, Karin. 1970. *Deutsche Kolonialherrschaft in Afrika: Wirtschaftsinteressen und Kolonialverwaltung in Kamerun vor 1914*. Zurich and Freiburg i.Br.: Atlantis.

Hochschild, Adam. 1998. *King Leopold's Ghost: A Story of Greed, Terror, and Heroism in Colonial Africa*. New York: Houghton Mifflin.

Hoffmann, Florian. 2007. *Okkupation und Militärverwaltung in Kamerun*. Vol. 2. Göttingen: Cuvilier.

Jäckel, Herbert. 1909. *Die Landgesellschaften in den Deutschen Schutzgebieten: Denkschrift zur kolonialen Landfrage*. Jena: Fischer.

Kaeselitz, Rudi. 1968. "Kolonialeroberung und Widerstandskampf in Südkamerun (1884–1907)." In *Kamerun unter deutscher Kolonialherrschaft: Studien, vol.* 2, edited by Helmuth Stoecker, 11–54. Berlin: Rütten & Loening.

Koch, C.W.H. 1913. "Die Stämme des Bezirks Molundu in sprachlicher, geschichtlicher und völkerkundlicher Beziehung." *Baessler-Archiv* 3: 257–312.

Laburthe-Tolra, Philippe. 1981. *Les seigneurs de la forêt. Essai sur le passé historique, l'organisation sociale et les normes éthiques des anciens Beti du Cameroun*. Paris: Karthala.

Mandeng, Patrice. 1973. *Auswirkungen der deutschen Kolonialherrschaft in Kamerun: Die Arbeitskräftebeschaffung in den Südbezirken Kameruns während der deutschen Kolonialherrschaft 1884–1914*. Hamburg: Buske.

Oestermann, Tristan, and Peter Geschiere. Forthcoming. "Coercion or Trade? Self-Realization During the Rubber Boom in German *Kamerun* (1899–1913)." In *Beyond the Margins: The Political Economy of Life in Modern Africa*, edited by Wale Adebanwi. Bloomington: Indiana University Press.

Omosini, Olufemi. 1979. "The Rubber Export Trade in Ibadan: Colonial Innovation or Rubber Economy." *Journal of the Historical Society of Nigeria* 10: 21–46.
Osborne, Emily L. 2004. "'Rubber Fever': Commerce and French Colonial Rule in Upper Guinée, 1890–1913." *Journal of African History* 45: 44–65.
Petersen, Erich R. 1939. *Gummi aus Kamerun*. Berlin: Safari.
Preuss, Paul. 1899. "Über Ausnutzung von Kautschukpflanzen in Kamerun." *Tropenpflanzer* 3: 15–20.
Raumolin, Jussi. 1984. "L'homme et la destruction des ressources naturelles: La Raubwirtschaft au tournant du siècle." *Annales E.S.C.* 39: 798–819.
Rockel, Stephen J. 2006. *Carriers of Culture: Labor on the Road in Nineteenth-Century East Africa*. Portsmouth: Heinemann.
Rollins, William H. 1999. "Imperial Shades of Green: Conservation and Environmental Chauvinism in the German Colonial Project." *German Studies Review* 22: 187–213.
Rudin, Harry R. 1938. *Germans in the Cameroons 1884–1914: A Case Study in Modern Imperialism*. New Haven: Yale University Press.
Rüger, Adolf. 1960. "Die Entstehung und Lage der Arbeiterklasse unter dem deutschen Kolonialregime in Kamerun (1895–1905)." In *Kamerun unter deutscher Kolonialherrschaft: Studien vol. 1*, edited by Helmuth Stoecker, 149–242. Berlin: Rütten & Loening.
Rupp, Stephanie. 2011. *Forests of Belonging: Identities, Ethnicities, and Stereotypes in the Congo River Basin*. Seattle and London: University of Washington Press.
Schlechter, Rudolf. 1900a. "Kautschuk-Expedition nach Westafrika." *Tropenpflanzer* 4: 109–20.
———. 1900b. *Westafrikanische Kautschuk-Expedition*. Berlin: Kolonialwirtschaftliches Komitee.
Schulte im Hofe, A. 1906. "Zum Kautschuk-Raubbau in Kamerun." *Deutsche Kolonialzeitung* (July 28).
Stein, Ludwig von. 1902. "Bericht des Oberleutnants Frhrn. v. Stein über die Bertua-Expedition." *Deutsches Kolonialblatt* 18: 440–3.
Treue, Wilhelm. 1955. *Gummi in Deutschland. Die deutsche Kautschukversorgung und Gummi-Industrie im Rahmen weltwirtschaftlicher Entwicklungen*. Munich: Bruckmann.
Tropenpflanzer. 1898. "Der Kautschukkonsum im Verhältnis zur Produktion." 2: 78–9.
Tully, John A. 2011. *The Devil's Milk: A Social History of Rubber*. New York: Monthly Review Press.
Vansina, Jan. 1990. *Paths in the Rainforests: Toward a History of Political Tradition in Equatorial Africa*. Madison: University of Wisconsin Press.
Vos, Jelmer. 2009. "Of Stocks and Barter: John Holt and the Kongo Rubber Trade, 1906–1910." *Commodities of Empire. Working Paper* 12.
Warburg, Otto. 1897. "Kickxia Africana." *Tropenpflanzer* 1: 99–103.
———. 1900. "Denkschrift O. Warburgs zur Begründung der Kautschuk-Expedition." In *Westafrikanische Kautschuk-Expedition*, edited by Rudolf Schlechter, 250–54. Berlin: Kolonialwirtschaftliches Komitee.
Weinstein, Barbara. 1983. *The Amazon Rubber Boom 1850–1920*. Stanford: Stanford University Press.
Whitford, H.N., and Alfred Anthony. 1926. *Rubber Production in Africa*. Washington: Government Printing Office.

Winkler, Hella. 1960. "Das Kameruner Proletariat 1906–1914." In *Kamerun unter deutscher Kolonialherrschaft: Studien vol. 1*, edited by Helmuth Stoecker, 243–86. Berlin: Rütten & Loening.
Wirz, Albert. 1972. *Vom Sklavenhandel zum kolonialen Handel: Wirtschaftsräume und Wirtschaftsformen in Kamerun vor 1914*. Zurich and Freiburg i. Br.: Atlantis.
Zouya Mimbang, Lucie. 2013. *L'Est-Cameroun de 1905 à 1960: De la mise en valeur à la marginalization*. Paris: Harmattan.

5 Useful Knowledge

The Monetary Education of Children and the Moralization of Productivity in the Nineteenth Century

Sandra Maß

The monetary education of children and young people has gained more public attention since the emergence of the current financial and banking crises. However, the question as to what extent children should have contact with money and learn monetary practices has been a contested issue for more than two hundred years. While it was not until the eighteenth century that the economic education of children found its way into public discourse and social practices in Western Europe and Britain, the discussion intensified at the end of the nineteenth century, above all, in connection with social and moral concerns. With the help of didactic material such as text and conduct books, board games, children's literature, and account books, the overriding desire was for children to be trained within school and family environments and made familiar with those practices. Questions of character and emotion were just as important as the physical dimension of monetary actions (Maß, forthcoming; see e.g., Frevert et al. 2014).

This chapter will adhere to recent scholarship on capitalism that underlines the cultural dimension in the construction of the economic sphere (see e.g., Dejung et al. 2012; Neu 2013; Dejung et al. 2014). Capitalism can be regarded as both a cultural and an economic system, as has been stated by Joyce Appleby (2011, 4), among others.[1] If we consider capitalism not only to be a structure that organizes economic activity by monetary exchange of values and profits but also as a model of social organization for people who use money in their daily lives, then models of monetary conduct, capitalist subject formation, and, particularly, the rehearsal of monetary practices become essential for the maintenance of this order. In this sense, children and young people are a promising object of analysis, because they typify the potential openness and precariousness of the education endeavor.

The example of monetary education in nineteenth-century Europe makes clear that children were considered a future resource and potential economic citizens. This becomes evident when looking at the middle-class discourse on moderation. In the course of the eighteenth and nineteenth centuries, waste and greed evolved from deadly sins into the main issues of bourgeois society and educational concern in Western Europe. Against this background, monetary education targeted both the body and the emotions, while using the topos of self-control to underline the necessity of moderation (Kessel 2000). This training was based on three intertwining elements: the construction and bodily control of needs, learning and training how to use money adequately, and, finally, acquiring knowledge about money. Theoretically, this process can be classified as embodiment. Body, cognition, and emotions were combined in most elements of monetary education: the body of the child had to learn and train to use money in certain ways and to experience the appropriate feeling when confronted with money and needs, and the child had to play and read in order to strengthen this process.

At the beginning of the period under examination, productivity was predominantly viewed in terms of the children's prospects for the future. It was envisioned that both girls and boys would be able to make ends meet, find a balance between waste and greed, and, finally, accept their appropriate place in society relative to class and gender.[2] The formation and control of productive bodies and their emotions was seen as the main goal (Eitler and Scheer 2009). Yet this notion changed gradually over the course of the nineteenth century. The first section of this chapter will briefly focus on the ideas that exponents of political economy held about passions at the beginning of the nineteenth century. The second section will demonstrate how teachers and school headmasters popularized some of these ideas and how they imagined the relationship between children and money. The third section shifts the focus to the material culture of children and analyzes how the benevolent and the consumptive child were constructed and how both were to be educated. Based on a wide range of sources such as children's literature, youth magazines, and toys, this chapter will present the argument that the concept of a productive child was very different at the end of the nineteenth century compared to that of the beginning of the century. Whereas benevolence was still important as a monetary habitus, it was increasingly supplemented by consumption. This shift accompanied the general historical change in moral orientation over the course of the nineteenth century: Whereas philanthropy and benevolence remained important in the construction of the middle-class *Wertehimmel* (value system), consumption became more and more important as an expression of habitus and an everyday necessity in the late nineteenth century.

Political Economy and the Control of Passions

The early modern assumption that human beings are marked by basic passions that balance each other had lost its interpretative power in the eighteenth century. However, in debates on the constitution of man, this belief was still adhered to in the presupposition that man was characterized by strong, internal driving forces. The designations and concepts of these driving forces—as feelings, passions, addictions, or impulses—varied considerably.[3] Yet commentators viewed them as potentially dangerous characteristics of the human species that must be brought under control. Appeals to appropriately guide the inner desires of the population were directed either at the government, and mediated by state institutions such as schools or prisons, or at the family where self-control and self-guidance were expected to be taught and practiced. The control, containment, and transformation of harmful passions, such as greed or prodigality, were declared politically and humanly necessary. Men and women who were led by harmful passions were adjudged to be a burden on local poor relief and a threat to societal morality. The general discourse about human passions was not purely economic: it also took place within the realm of the religious, philosophical, and pedagogical debates about the human condition. The German pedagogue August Herrmann Niemeyer (1754–1828), for example, described selfishness as a curse that knows different subforms: envy, acquisitiveness, avarice, greed, and monetary love (Niemeyer [1796] 1970, 192).

Among the general discourse about passions, there was one group of authors with an extraordinary interest in connecting the question of passions with the realm of the economic. The rising political economy in Great Britain and France and, to a somewhat different degree, in the German-speaking territories encouraged this group to comment on the general state of education and, thereby, identify unconstrained passions as a major obstacle to an ordered economy. Adam Smith (1723–1790), for instance, wrote in *An Inquiry into the Nature and Wealth of Nations* about the necessity to implement an education adapted to the economic system. He was concerned with the negative effects of the division of labor that he praised as the foundation of a prosperous economy at the same time (see the introduction to this volume). According to Smith, the division of labor meant that working men performed only one or two actions and, as a result, gradually atrophied: "He naturally loses, therefore, the habit of such exertion, and generally becomes as stupid and ignorant as it is possible for a human creature to become" (Smith [1776] 1786, 2: 182). As a corrective, he underlined the importance of education for the existence of human kind. Thus, influenced by pedagogics and other enlightened ideas, Smith and others declared that societal progress by dint of economic growth, a progressive state, and a relatively comfortable situation of the laboring classes could only be achieved through the improved education of children and young people (Redman 1997, 139).

However, at that time, common people's children had little opportunity to take care of education:

> They have little time to spare for education. Their parents can scarce afford to maintain them even in infancy. As soon as they are able to work, they must apply to some trade by which they can earn their subsistence.
>
> (Smith [1776] 1786, 3: 186–7)

He did not suggest a general prohibition of child labor. Yet his demands encompassed an expansion of local elementary schools, the parish school:

> But though the common people cannot, in any civilized society, be so well instructed as people of some rank and fortune, the most essential parts of education, however, to read, write, and account, can be acquired at so early a period of life, that the greater part even of those who are to be bred to the lowest occupations, have time to acquire them before they can be employed in those occupations.

After 1800, nearly all political economists embraced Smith's position. Their argument that it was cheaper to offer a minimal school education than to provide for the increasing amount of criminals convinced the last critics (West 1964). For instance, the German professor for *Nationalökonomie*, Karl Heinrich Rau (1792–1870), underlined that the extension of the elementary school system would increase the "moral and spiritual condition of a people" and help them to understand trade and to maintain "order in family households" (Rau 1828, 24). Education would prevent poverty and have long-term effects in fostering the "rule of quiet intellect over passion" (379–80), especially in questions of early marriage and family foundation.[4] Avoidance of criminality, transfer of basic knowledge, and control of passions were regarded as the basis of the social and economic order, thus allowing for regular work within a profit-oriented economy, the reduction of (unproductive) poor relief, and stable economic growth.

In this vein, the British utilitarian James Mill (1773–1836) in his *Elements of Political Economy* pleaded for a "generous and animating diet" leading to a moderation of passions:

> A perfect command, then, over a man's appetites and desires; the power of restraining them whenever they lead in a hurtful direction; that possession of himself which insures his judgment against the illusions of the passions, and enables him to pursue constantly what he deliberately approves, is indispensably requisite to enable him to produce the greatest possible quantity of happiness.
>
> (Mill [1820] 1825, 29, 15)

According to Mill, all means of education were aimed at one goal: The human mind should facilitate happiness. As he put it in the preface, his textbook was written for both sexes and members of all classes. He underlined the general necessity of economic education for both girls and boys, while, at the same time, offering differentiated contents. The same holds true for the different classes: Economic knowledge and monetary expertise were supposed to be part of the basic equipment of all, the middle class as well as the working class and the rural youth. However, he differentiated between liberal economic ideas based on private property, free trade, and capital formation, on the one hand, and more traditional conceptions of careful spending, social immobility, and peasant husbandry on the other. Like Mill, Jane Marcet (1769–1858), author of *Conversations on Political Economy*, a book that set out to popularize political economy especially for girls and women, conceived economic theory as an explicit antagonist of the human passions: "Political economy is particularly inimical to the envious, jealous, and malignant passions" (Marcet 1816, 20).

Regardless of their position toward political economy—whether teaching it was considered to be helpful for controlling the human passions or not—these authors were certain that they could not ignore the dangers that confronted man's emotional order within capitalist societies. Special attention was directed toward those financial dealings that were classified as particularly risky: the money-related passions of greed and waste. The German-Russian economist Heinrich Storch (1766–1835), for instance, described the relationship between these passions and money as tight and dangerous:

> The consumption of the assets is always connected with the affections and passions of men . . . . To pursue the path of wisdom is, as everywhere, the most difficult . . . . The errors of excess in using the assets are waste and avarice.
>
> (Storch 1819, 2:)

Avarice was seen by Storch as a "drive, a mechanical, dishonoring need" that was directed at a future that would never happen (166). Ultimately, argued Storch, avarice could only end in death. This prospect made it an even greater evil than waste, as the latter was at least regarded as social and would have its absolute limits in scarcity (195).

Children and Money

When political economists focused on education and the control of passions, they rarely touched directly on the monetary education of children. However, members of the educational system popularized their ideas, transformed them into school practices, and expanded upon them. Albert Gerth, for example, a German teacher at the royal Pädagogium at Rügen directed his 1839 book *Buchhaltung für Kinder* (*Accounting for Children*) at boys who were about to leave their family to study or apprentice. The author underlined how important the small-scale order was for the viability of the state economy. Children should, therefore, he wrote, be provided with economic knowledge in school:

> And if the school should be a place of preparation for the boy, its rules and warnings educating him morally and intellectually to a future man and citizen, why is the school not offering direct advice on how to budget the material conditions of his coming middle-class (*bürgerlich*)

> existence? Why is it not giving him certain rules of economy and thrift to take along?
>
> (Gerth 1839, V–VI)

At the same time, he presented another argument in favor of monetary education that was commonly used among contemporaries. Most cases of waste in adulthood, he wrote, go back to a wasteful handling of pocket money during childhood (VIII): "A boy seduced to disorder, waste, and indulgence will never achieve anything proficient as a student and even less will he meet the requirements of his future position as a citizen within a state" (IX–X). Thus, Gerth thought of the state order as being based on the economic reason of men, rather than referring to political economy in which order was based on the assumption that scarcity would be regulated through the supply and demand of the market.

Integrating political economy, in particular, and economic knowledge, more generally, into school curricula was the main target of Benjamin Templar (1831–1879). The headmaster from Manchester directed his economic schoolbook *Reading Lessons in Social Economy* at male readers from nine years of age onward. His book dealt with elementary presentations of certain aspects of contemporary political economy, like the division of labor, profit, capital, and money. The latter was seen as the general means of exchange:

> All boys who now go to school, know that the time will come, when they will go to work, to earn their own "living", as some say; or, what means the same thing, to get wages. With the money they get as wages, they or their parents will buy food, clothes, and other things they want. Money itself is of no use; people can neither eat, drink, nor wear it, nor can they live in it; and yet they work very hard to get it. The reason why they do so, is this; to live, they must have food, clothes, fuel, and shelter;—that is, places to live in. Money, and money only, will get them these things; and therefore it is, that people work to get it. We see then, that although to get money, is what all work for, money is not the real object of their labour, but what money will buy.
>
> (Templar 1858, 1)

Templar took up the tradition of public enlightenment and designed his instructions with the help of moral stories. Like others before him, he divided the world into good and bad dealings with money. To this end, he started by putting forward a negative example. An anonymous worker who, at the beginning of a month, did not think about its end, wasted his entire wage with extravagance:

> Such a man can have but very few of the comforts of life, not because he has not money enough to buy them, but because he does not manage

> well with what he has;—he is a bad manager, and bad management is the cause of his poverty and discomfort.
>
> (Templar 1858, 13)

To contrast this story of failure, he gave a positive counterexample: Smith, the good worker, organized it all well, always saved money, and was able to live more decently.

> To live comfortably, he must not only possess the knowledge, skill, and industry, that are necessary to get a good wage, but he must have good management. Good management is called economy; and a person who is a good manager, who possesses economy, is said to be an economical person.
>
> (14–15)

For Templar—as for many other contemporaries in Britain and in the German-speaking lands—reasonable economic behavior was based on thrift and the knowledge of when to spend the saved money. Thrift was regarded as the economic behavior that would prevent the flaws of waste or avarice. It was defined as a means of moderation, as a balance between two passionate poles.[5] Thus, in its popularized versions, such as in schoolbooks, political economy was boiled down to teaching the meaning of money and the practice of thrifty behavior. As such, this training was not too different in content from the information spread in educational manuals and conduct books by the churches and by families.[6] Yet, whereas some British writers tried to integrate the ideas of political economy into schoolbooks, the German case is slightly different: German economists also pleaded for an education that was related to the economic situation. However, their ideas remained limited and did not integrate political economy as such.

Benevolence and Consumption as Signs of Children's Productivity

Whereas some political economists, such as James Mill with his son John Stuart Mill, tested the practical effects of their educational ideas directly at home, it was up to education manuals, toys, and children's literature to suggest practical exercises for a broader audience.[7] On all three levels—manuals, toys, and children's literature—the habitus of benevolence, thrift, and appropriate consumption dominates the money-related educational goals over the course of the nineteenth century. However, the emphasis shifted from benevolence in the first half to consumption in the second half of the century. Throughout the century, the basis of charity and consumption by middle-class families was seen to be the possession of and the control over money. In the case of working-class children, money was mostly earned with great difficulty, begged for on the streets, and used to make ends meet. In middle-class families, the gift of pocket money gained increasing acceptance and became a widespread practice in the second half of the nineteenth century. An allowance was seen as a way of learning both social and monetary behavior and, thus, should not exceed the limits of the family finances or the accepted norms of the respective social order. Children were given regular pocket money on the condition that they did not spend it immediately and that they saved it appropriately. This led to the creation of suitable storage receptacles. Sources from middle-class, rural, and working-class families show that the money-box was widespread among children of all classes over the course of the second half of the nineteenth century (see Thurn 1982; Felder 1904, 23).

Until well into this period, pocket money was to serve primarily as a means to allow children to practice charity. The benevolent gift of money to paupers and the donation of funds for the Christian missions were omnipresent as normal, social practices in a child's socialization. The donation was not only valued as an ease for the poor, but it was also supposed to produce contentment and well-being for the donor. Poor children and Africans were especially deemed needy and described as the legitimate recipients of the donated pocket money. "How many negroes, orphans, and other poor children might then be benefitted", noted a reverend of a Sunday school in 1829.[8] A short story popular at the time told of a boy who hears a coin speak and says, "You shall be a lovely keepsake and a reminder to be benevolent towards poor people and needy people and also towards pagan children in far-off lands" (Frankenberg 1886, 111). The literary model for the benevolent child was the fairy-tale figure of *Sterntaler* (The Star Money), a girl who owns very little and gives away what she has when she encounters someone needier. *Sterntaler* was presented as the counterimage to the wasteful boy and the mean, wealthy man. She would be rewarded in the end of the tale with endless money given to her because of her empathy and altruism.

In children's literature, the act of donating was unanimously described as important, but for different reasons. Especially in Britain, philanthropic behavior was seen as a replacement for public assistance (Ritter 1989, 45–7). It was supposed to fulfill a self-help function, enabling the poor to regain the ability to work and sustain their families. Other examples of benevolent behavior served to demonstrate the limits of wealth and the importance of moral education. Therefore, descriptions of benevolence were based on the suppression of immediate needs. The desire to purchase sweets or firecrackers should be given up in order to be able to drop a couple of coins into the collection in Sunday schools or churches. Thus, not only the situation of the pauper was meant to be improved by the donation but also the character of the wealthy child, who could prove his or her devotion to the Christian concept of charity through the donation. The stories and narratives about benevolent children all followed a similar script: Children's books relentlessly repeated stories of embarrassment when a child—generally a boy—could not donate any money on Sunday because he had already wasted it.

The implicit construction of unproductive children, in the sense that they had not suppressed their immediate needs in favor of a more worthy, future cause, was one of the key elements of how productivity became relevant as a moral issue in children's monetary education.

The figure of the thriftless boy was expanded upon in children's literature by descriptions of wealthy people who hoarded their money, despite the fact that they owned considerable assets, and who rifled through the possessions of those who had less or nothing. Immoderate wealth without care for those who owned less was regarded as negative and seen as potentially dangerous for the character: "They are deaf to charitable efforts, to benevolence, and they stop more or less being useful members of society" (Das Pfennig-Magazin für Kinder 1837, 389). Normally, the story ends badly for such wealthy but corrupt characters: They lose their money, they get sick, or die, as in a British short story published around 1840 (Ospringe [1840], 11: 28).[9] The author described how a couple of boys, aged between ten and fifteen, discussed the search for the "most avaricious man in history" at a club in Warwickshire. After the debate, they listened to a presentation of an elder member of the club about an old-established family. The patriarch had rented out his houses expensively to poor people and had even taken away the gardens that had guaranteed their subsistence. He ignored the admonition of the local clergyman and, subsequently, became the victim of a robbery committed by young people angry about his repeatedly mistreating a poor female tenant. All his money was stolen, and the loss made him ill. Yet, as he was now unable to pay the doctor, he ended up in debt and was taken to court. The worry eventually killed him.

Although only a few of these short stories went so far as to present the outcome of avarice to be death, children's materials, in general, underlined the dire consequences when wealthy people would not behave benevolently. Owning money came with moral duties to the community in the first half of the nineteenth century. In the second half of the century, however, more pragmatic voices in children's and youth media supplemented this discourse. Increasingly, the ideal of reasonable consumption was presented as a different way of spending money and as an expression of another mode of productivity, which was connected to the investment of money that would generate growth in the course of the business cycle. Donating money was still seen as the source of "real happiness", but the emphasis gradually shifted, and consumption became an accepted way for children to use their money if it happened in a controlled manner. Previously, the child's capacity to make a charitable gift had been considered as having the desired character-building and instructive effect. Now, educational efforts were increasingly directed toward the awareness of and ability to cope with monetary dangers in the consumer society. Thus, the focus of children's media shifted to spending money on consumer goods and gaining funds as a means to achieve independence.[10]

The concept of productivity that was implicitly connected with children's acts of consumption lay somewhat at odds with the idea of the benevolent child. Now, productivity was seen as the result of a successful education process through which children should learn the value of the respective coins, the exchange of goods and money, and the dangers of the possession of money. Whereas the construction of the consuming and money-spending child, thus, differed from the idea of a philanthropic character (with all the various layers of productivity hidden within the idea of benevolence), there are still parallels to be drawn between the two regimes. The consuming child should also know how to use money, save and spend it at the right time, and behave reasonably as a citizen of a—albeit commercialized and financialized—society. Using the perspective of productivity, however, reveals a different emphasis that lay behind the figure of the consuming child: spending money for goods was an act of investment for an expected output, for an increase in production cycles, and an increase in profit making.

Having said this, it is slightly surprising that the economic influence was not overly present in the realm of games, toys, and books. Images associated with money only played a minor role, alongside images of geography and nature found in puzzles or parlor games. Yet one of the most popular jigsaw puzzles, *The Pence Table* (1790), was an instruction on the translation of units of money, from pence to shillings and vice versa (Hannas 1972, 108–09). Jigsaw puzzles were often illustrated. The puzzle, *Poor Richard, by Benjamin Franklin* (1791), for example, showed a quotation by Franklin that had to be pieced together:

> At this time when the crying complaint is that money is so scarce, it must be an act of kindness, to inform the moneyless, how they can reinforce their purses. I will acquaint you with the true secret, of money catching, the certain way to fill empty pockets, and how to keep them always full.
>
> (Hannas 1972, 114)[11]

Spending money beyond simply making donations also became a topic in children's literature. In the short story *Charlie's Sovereign, and How He Spent It*, using money for immediate consumption is depicted as an acceptable boyish behavior in contrast to the undesirable hoarding of money (The Boy's Yearly Book for 1865, 226):

> When he finds, after the parting hand-shake, that a medallion of good Queen Victoria has been left in the palm of his hand, in the shape of a coin of the realm. Be the tip a golden sovereign or even half-a-crown, according to the age of the recipient or the purse of the giver, it is very acceptable, and the spending of it forms a pleasing interlude in the monotony of life at school. Certainly, we have known boys who did not spend their tips, boys to whom the parting with money, either for books or less endurable sources of satisfaction, possessed no charm; but we are glad to believe, for the credit of the boyish nature, that such instances are rare.

Training to consume became an unquestioned necessity in discourses about education. Additionally, gendered elements entered the general assumption that money had become more important and everybody should, therefore, learn how to use it. With the legal changes in English property rights in the 1880s, girls and women were allowed to own money even after marriage. In 1885, one author argued in *The Girls Own Paper* that the changes in property law produced the need for training in the use of money:

> Every girl, to my thinking, ought to have an allowance for dress and her own small personal expenses proportional to her position in life, for it is the only way by which they will realize that money will do, how far it will go, and how much art there is in the manner of spending it.

It would be advantageous, the author continued, to keep an account book and to pay immediately for the purchases: "It is a duty in life, and a great one too, to be very prompt and particular about money payments, and the well-to-do have duties as well as the needy" (Holt 1885, 167).[12] Secular magazines for girls underlined the importance of money, as depicted in another article from 1886:

> A great part of the unhappiness of the world is caused by carelessness about money. Money is no trifle. Without it, we cannot lead rational lives, or ever be really independent. We must not set our hearts on it, but we must recognise its true value, and know wisely how to use it.
> (Mason 1890, 167)

Happiness would result from the right use of money and independence and rationality were necessary to live a good life, the author continued.

At the end of the nineteenth century, children's and young people's magazines increasingly formulated advice that surpassed the familiar references to female household management. Buying, selling, profit, and loss were of great interest for the human spirit in general, and the child's living environment, in particular, according to another English magazine article about girls and their pocket money from 1897 (*The Girl's Own Paper* 1897, 418). Interestingly, the notion of productivity was not visible in this article. Yet, in 1886, *The Girl's Own Paper* published a series of articles titled "Every Girl a Businesswoman" by James Mason, in which he took the new legal and economic circumstances into account and stressed that it should be the goal of modern girls to become businesswomen. Far from being contrary to the female mind, Mason (1886, 5) argued that knowledge of money and independent economic action were no longer purely male domains:

> Every girl who is guided by common sense will aim at becoming a business woman. That is to say, she will try to cultivate habits of order, industry, perseverance, method and punctuality, and will do her best

> to learn how to conduct formal correspondence, how to keep account, how to manage money, and what to do with savings. Beside this, she will make a point of knowing something about the laws relating to domestic life—the renting of houses and the employment of servants, for example . . . Perhaps someone may say that "business woman" has a hard sound, and stands for a character precise, selfish, and uninteresting. That is not what we intend by it at all. Is a girl to be less lovable, less gentle, less charming, whenever we cease her to say of her, That [sic] girl, in regard to all the ways of business, is a perfect simpleton? On the contrary, business is a fine training-school for many virtues; and of all good women, a good business woman may be reckoned the very best.

In addition to advice for older girls who were taught about the economy and money, such texts offered practical and playful training on general matters of consumption for younger children. The economic sphere, in general, and consumption, in particular, became even more important with the spread of play money and play shops. The producers never directly named any educational goals nor were they explicitly talking about productivity, but play shops and play money allowed children to simulate the act of everyday consumption and trained them for their future economic life as adults. Seen from the perspective of the historian, one could ask whether the playful imitation of trading, selling, and exchanging money was designed to educate children in productive behavior.

Replicas of shops had evolved from dollhouses, which became more popular and more affordable since the end of the eighteenth century, whose function lay in the playful imitation of bourgeois family life. The dollhouses generally lacked any reference to the world outside the home or to the economy. Whereas they represented the inner world of the bourgeois home, play shops represented the outer world, particularly the world of trade. Dollhouses were originally individual productions for wealthy households rather than mass-produced goods. However, they became increasingly popular after they were produced industrially and, subsequently, became less expensive in the late nineteenth century. Working-class and artisan families could also afford a play shop by that time (Bachmann and Metzger 1992, 19).

The play shops tended to have a relatively uniform design. The rear walls were devoted to the presentation of the goods, and, in most cases, the shop included a counter, frequently with some form of checkout option to accept payment. This may have been a slit in the middle of the sales counter into which one could a toss in a coin, a cash box, or even a cash register; these items were produced as shop accessories after the turn of the century (Bachmann and Metzger 1992, 66). Moreover, the gradual integration of payment facilities into the play shop in the last third of the nineteenth century can also be traced to play money collections (Aschoff 2009). Play money has long been used as an object of play known, for example, as toy coins, *Rechenpfennig*, or tokens (Rogers 1990, 23). Whereas many play money sets displayed the respective insignia and imitated the official currency of the respective country, there were others sets upon which animals or fantasy characters such as the *Struwwelpeter* were depicted (26). The coins were sold individually or in a cash box. Children's money could be stored and exchanged and, thus, function as an imitation of the real means of exchange. However, children decided for themselves how to use the coins in their games.

Children's lives were far more than the sum of moral literature and attempts to educate them and, of course, consisted of more than monetary exchange and donations. However, the actual use and interpretation of money by children has rare evidence in primary sources. Nevertheless, autobiographical sources offer some insights into the world of children (Hopkins 1994, 312–13; Jarzebowski and Safley 2014). Documents from people with working-class backgrounds show that the street was the dominant space for children's everyday experiences. It was the place where they worked, had contact with money, and engaged in daily begging, stealing, and interacting with products and prices (Hopkins 1994, 194–5; Davin 1996). Thus, the street had little in common with the imitation of shopping constructed by a play shop placed in the middle-class living room.

At the same time, middle-class children used the street only in company with other people. Thus, the street became attractive as a space of immediacy in which the norms and morals of the middle class seemed to be absent. In childhood memories, the street is given a prominent place only in exceptional cases. However, middle-class children occasionally documented attempts to leave their class-specific spaces of socialization—the house and the school. For instance, Gerhart Hauptmann (1862–1946), the German author and playwright whose parents ran a spa hotel, depicted himself as crossing between the bourgeois home and the proletarian street. He described his attempt to be a worker on the street, carrying the luggage of the spa guests and getting paid for it. He confessed that he did it only once because the payment was too low, the bags too heavy, and the treatment too bad (Hauptmann 2007, 33). His coquettish description of his normally childish play and his, albeit brief, experience with being a hardworking street boy were shaped by the knowledge that he was not forced to do what he did. In the end, it was not the poverty of the street boys that constituted his monetary socialization, but the contact with the wealthy spa guests (38).[13] Hauptmann's example reveals that historians should be cautious when taking manuals, didactic toys, schoolbooks, and other materials to be adequate representations of what the world of children in the nineteenth century looked like. The attraction of "the other world", as it is depicted in Hauptmann's autobiography, hints at a place beyond the educational efforts and impressions of the economically active and, for the larger part at least, productive citizens.

Conclusion

When considering children and productivity in the nineteenth century, child labor immediately comes to mind (Humphries 2010). However, ideas and practices of monetary education are another perspective on children's productivity—subjects that have not yet gained the same scholarly attention. These ideas and practices can be traced from the intellectual world of political economists and the teaching concepts of headmasters to toys, children's literature, and youth magazines. Although the protagonists and materials differed considerably, their content was largely analogous: control of passions; avoidance of monetary passions, such as greed and waste in particular; and practical training in the use of money by donations and consumption. Generally, monetary education of all kinds centered on the middle-class concept of moderation and tended to be based on comparable approaches to money-related passions.

Although the very notion of productivity was not used, it was present implicitly by idealizing the children's futures in which they would use money reasonably, work and budget in their respective gendered spaces, learn how to consume, and, thus, become productive citizens. By the end of the nineteenth century, the old habit of donating money to the poor and needy gradually lost its importance in the discourse on monetary education. The right way to consume entered the world of children and became a central element of instruction for girls and boys alike. The history of monetary education, thus, offers an example of the change of emphasis in the creation of the productive forces of tomorrow from usefulness to productivity and consumption.

The useful member of society as constructed in the discourses on charity was more closely related to the idea of a morally respectable person, acting for the good of the respective community, whereas the productive member of the consumer society was portrayed as somebody who invests her or his money appropriately for the purpose of economic growth. However, it would not be adequate to speak of a process in which one discourse fully replaced the other. Elements of the discourse of the useful and morally acceptable child remained important and present in debates about education in general and consumption in particular. Far from being a sharply defined concept based on numbers and workload, productivity in discourses about monetary education consisted of various discursive elements, such as inappropriate emotions, the ideal of self-help and financial autonomy, and right ways of consumption. Thus, all kinds of children's dealings with money were related to questions of morality in two ways: First, they could behave immorally by not using money in the appropriate way. Second, their behavior as a child was read as a sign of their envisaged future, pointing either to an immoral, namely, unproductive life as an adult, or a moral and, therefore, productive one. Hence, morals and productivity were intertwined.

Notes

1 Jürgen Kocka (2010, 13), for example, takes a different approach. He underlines that capitalism does have cultural and other noneconomic preconditions and effects, but does not think of it as a distinct cultural system.
2 Regarding the idea of producing "productive forces" in Friedrich List's *National System of Political Economy*, see the introduction to this volume.
3 Some concepts were still related to Thomasius's (1696) *Lasterschema*; others preferred the state to be responsible for control of the potential compulsions (see Newmark 2008). Regarding the changes to related concepts in the eighteenth and nineteenth centuries, see the classic study by Albert O. Hirschman (1977).
4 Other authors who explicitly mentioned the connection between schooling and the economy were Jean-Baptiste Say, Johan Friedrich Eusebius Lotz, and Friedrich Julius Heinrich von Soden.
5 Jan de Vries and others have underlined the changing relation between households, time, and work in the course of the long eighteenth century. According to Vries (2008, 10, 40–72), the availability of an ever-growing market of goods around 1800 led to an increase in working hours, the "industrious revolution". In addition to this argument, the popularity of thrift in the second half of the nineteenth century is central to the understanding of monetary education. Not spending surplus cash but setting a sum aside for future use was a significant existential decision for the majority of people. Lack of job security and the possibility of unforeseeable illness meant that making financial provision for the future was essential for the lower social classes. For the middle classes, however, pecuniary reserves were frequently the basis or result of an accumulation of capital that would enable further investment. Regardless of the various motives for thriftiness, Joshua J. Yates and James D. Hunter (2011, 5) described it recently as a social norm that constituted the "moral economy of the self". As an idea, a virtue, a central goal of education, and as a social practice, thriftiness was closely flanked by other virtues such as punctuality, order, respectability, and moderation and was a significant concept in an expanding culture of seeking and giving advice.
6 For a more detailed analysis of economic knowledge in different social and religious spaces, see Sandra Maß (forthcoming).
7 Regarding John Stuart Mill's description of his father's educational attempts, see John Stuart Mill (1957, 3–88). Regarding children as consumers in the nineteenth century, see Dennis Denisoff (2008).
8 Articles on charity were published on a regular basis in Christian child and youth magazines (see, for example, E.K. 1827, 156; W.A.R. 1829, 310; Bartlett 1882, 875–7).
9 The series consisted of thirteen volumes with numerous moral tales of approximately twenty pages each.
10 Children in eighteenth century England were already considered to be consumers, as J. H. Plumb (1982) has shown. However, the existence of these "child consumers" was limited to metropolises. The situation did not begin to change until the nineteenth century. In Germany, it was not until the last third of the nineteenth century that children's worlds became more generally commercialized (see Hamlin 2007).
11 Reliable statements about the dissemination of the games are difficult to make. Some were launched repeatedly over years or decades, which suggests a strong demand in bourgeois households.
12 Also, other youth magazines underlined the necessity of monetary knowledge (see, for example, Young Folks Paper 1886a, 26; 1886b, 371).
13 For further examples of autobiographical sources of monetary experiences in childhood, see Maß (forthcoming).

Bibliography

Appleby, Joyce. 2011. *The Relentless Revolution: A History of Capitalism.* New York: Norton.

Aschoff, Günter. 2009. *Deutsches Kinderspielgeld: Ein numismatisches Randgebiet.* Regenstauf: H. Gietl Verlag & Publikationsservice GmbH.

Bachmann, Manfred, and Wolfram Metzger, eds. 1992. *Vom Marktstand zum Supermarkt: Der Kaufladen in Puppenwelt und Wirklichkeit; Katalog zur Ausstellung des Badischen Landesmuseums im Schloss Bruchsal.* Karlsruhe: INFO.

Bartlett, G.B. 1882. "How the Children Earned Money for Charity." *St. Nicholas* (September 1): 875–7.

The Boy's Yearly Book for 1865. 1865. "Charlie's Sovereign, and How He Spent It." 226–9. London: S.O. Beeton.

Das Pfennig-Magazin für Kinder. 1837. "Der Geiz." (December 9): 389.

Davin, Anna. 1996. *Growing Up Poor: Home, School, and Street in London, 1870–1914.* London: Rivers Oram Press.

Dejung, Christof, Monika Dommann, and Daniel Speich Chassé, eds. 2012. "wissen und wirtschaften." *WerkstattGeschichte* 58 (special issue).

———. 2014. *Auf der Suche nach der Ökonomie: Historische Annäherungen.* Tübingen: Mohr Siebeck.

Denisoff, Dennis. 2008. "Small Change: The Consumerist Design of the Nineteenth-Century Child." In *The Nineteenth-Century Child and Consumer Culture*, edited by Dennis Denisoff, 1–26. Aldershot: Ashgate.

Eitler, Pascal, and Monique Scheer. 2009. "Emotionengeschichte als Körpergeschichte: Eine heuristische Perspektive auf religiöse Konversionen im 19. und 20. Jahrhundert." *Geschichte und Gesellschaft* 35: 282–313.

E.K. 1827. "A Penny Saved Is a Penny Gained." *The Child's Companion or Sunday Scholar's Reward* (May 1): 156.

Felder, Franz Michael. 1904. *Aus meinem Leben.* Wien: Verlag des literarischen Vereins.

Frankenberg, Paul. 1886. *Erlebnisse eines Marienthalers: Eine Erzählung für die liebe Jugend.* Mainz: Kirchheim.

Frevert, Ute, et al. eds. 2014. *Learning How to Feel: Children's Literature and Emotional Socialization, 1870–1970.* Oxford: Oxford University Press.

Gerth, Albert. 1839. *Buchhaltung für Kinder oder Anweisung zur Ordnung und Sparsamkeit in Geldsachen für den Selbstunterricht und Gebrauch in Schulen.* Stralsund: Löffler.

The Girl's Own Paper. 1897. "Girls and Their Pocket-Money." (April 3): 418.

Hamlin, David. 2007. *Work and Play: The Production and Consumption of Toys in Germany, 1870–1914.* Ann Arbor: The University of Michigan Press.

Hannas, Linda. 1972. *The English Jigsaw Puzzle, 1760 to 1890; With a Descriptive Check-List of Puzzles in the Museums of Great Britain and the Author's Collection.* London: Wayland.

Hauptmann, Gerhard. 2007. *Das Abenteuer meiner Jugend/Buch der Leidenschaft: Autobiographische Romane.* Berlin: Ullstein.

Hirschman, Albert O. 1977. *The Passions and the Interests: Political Arguments for Capitalism before Its Triumph.* Princeton, NJ: Princeton University Press.

Holt, Addern. 1885. "Money Obligations." *The Girl's Own Paper* (December 12): 167.

Hopkins, Eric. 1994. *Childhood Transformed: Working-Class Children in Nineteenth-Century England*. Manchester: Manchester University Press.
Humphries, Jane. 2010. *Childhood and Child Labour in the British Industrial Revolution*. Cambridge: Cambridge University Press.
Hunter, James D., and Joshua J. Yates. 2011. "Introduction: The Question of Thrift." In *Thrift and Thriving in America: Capitalism and Moral Order from the Puritans to the Present*, edited by James D. Hunter and Joshua J. Yates, 3–33. New York: Oxford University Press.
Jarzebowski, Claudia, and Thomas M. Safley, eds. 2014. *Childhood and Emotion Across Cultures, 1400–1750*. London: Routledge.
Kessel, Martina. 2000. "Das Trauma der Affektkontrolle. Zur Sehnsucht nach Gefühlen im 19. Jahrhundert." In *Emotionalität. Zur Geschichte der Gefühle*, edited by Claudia Benthien, Anne Fleig, and Ingrid Kasten, 156–77. Köln: Böhlau.
Kocka, Jürgen. 2010. "Writing the History of Capitalism." *Bulletin of the German Historical Institute Washington* 47: 7–24.
Maß, Sandra. Forthcoming. *Kinderstube des Kapitalismus: Monetäre Erziehung im 18. und 19. Jahrhundert*. München: De Gruyter Oldenbourg.
Marcet, Jane. 1816. *Conversations on Political Economy: In Which the Elements of that Science are Familiarly Explained*. London: Longman.
Mason, James. 1886. "Every Girl a Business Woman: A Practical Guide to the World of Industry and Thrift." *The Girl's Own Paper* (October 2): 5–7.
Mason, Nanette. 1890. "Schoolgirl Troubles, and How to Cope with It." *The Girl's Own Paper* (December 13): 166–7.
Mill, James. (1820) 1825. "On Education." In *Essays on Government, Jurisprudence, Liberty of the Press, Prisons and Prison Discipline, Colonies, Law of Nations, Education, Reprinted by the Permission from the Supplement of the Encyclopedia Britannica*, edited by James Mill, 1–46. London: J. Innes.
Mill, John Stuart. 1957. *Autobiography*. Indianapolis: Bobbs-Merrill.
Neu, Tim. 2013. "Symbolische Kommunikation und wirtschaftliches Handeln: Theoretische Perspektiven." In *Alles nur symbolisch? Bilanz und Perspektiven symbolischer Kommunikation*, edited by Barbara Sollberg-Rilinger, Tim Neu, and Christina Brauner, 401–18. Köln: Böhlau.
Newmark, Catherine. 2008. *Passion—Affekt—Gefühl: Philosophische Theorien der Emotionen zwischen Aristoteles und Kant*. Hamburg: Meiner.
Niemeyer, August Hermann. (1796) 1970. *Grundsätze zur Erziehung und des Unterrichts für Eltern, Hauslehrer und Erzieher*, edited by Hans-Herrmann Groothoff and Ulrich Herrmann. Paderborn: Schöningh.
Ospringe, Emily. [1840?]. *Parlour Stories: A Series of Interesting Tales for the Amusement and Instruction of the Young*. Vol. 11. London: [Edward Lacey?].
Plumb, J.H. 1982. "The New World of Children in Eighteenth-Century England." In *The Birth of a Consumer Society: The Commercialization of Eighteenth-Century England*, edited by Neil McKendrick, John Brewer, and J.H. Plumb, 286–315. London: Europa.
Rau, Karl Heinrich. 1828. *Grundsätze der Volkswirthschaftspflege mit anhaltender Rücksicht auf bestehende Staatseinrichtungen*. Heidelberg: Winter.
Redman, Deborah A. 1997. *The Rise of Political Economy as a Science: Methodology and the Classical Economists*. Cambridge: MIT Press.
Ritter, Gerhard A. 1989. *Der Sozialstaat. Entstehung und Entwicklung im internationalen Vergleich*. München: Oldenbourg.

Rogers, David J. De Sola. 1990. *Toy Coins*. Wolverhampton: Galata.
Smith, Adam. (1776) 1786. *An Inquiry into the Nature and Causes of the Wealth of Nations*. Vol. 2, 3. London: A. Strahan and T. Cadell.
Storch, Heinrich. 1819. *Handbuch der National-Wirtschaftslehre*. Vol. 2, edited by Karl Heinrich Rau. Hamburg: Perthes und Besser.
Templar, Benjamin. 1858. *Reading Lessons in Social Economy for the Use of Schools*. London: Jarrold and Sons.
Thomasius, Christian. 1696. *Von der Artzeney wider die unvernunfftige Liebe und der zuvorher nöthigen Erkäntnuß sein Selbst, oder: Ausübung der Sittenlehre*. Halle: Salfeld.
Thurn, Hans Peter. 1982. *Die Kultur der Sparsamkeit: Die Kulturgeschichte des Sparens: Illustriert mit Spardosen aus zweitausend Jahren*. Stuttgart: Deutscher Sparkassenverlag.
Vries, Jan de, 2008. *The Industrious Revolution: Consumer Behaviour and the Household Economy, 1650 to the Present*. Cambridge: Cambridge University Press.
W.A.R. 1829. "On Spending Money." *The Child's Companion or Sunday Scholar's Reward* (October 1): 310.
West, E.G. 1964. "The Role of Education in Nineteenth-Century Doctrines of Political Economy." *British Journal of Educational Studies* 12: 161–72.
Young Folks Paper. 1886a. "Spending Money." (January 9): 26.
———. 1886b. "How to Value Money." (June 12): 371.

6 The Contested Productivity of the Baker's Body

Technology, Industrialization, and Labor in Nineteenth Century France

François Jarrige

With thirty-three thousand shops recorded in 2011—one per two thousand inhabitants—France has the highest density of bakeries in the world.[1] Independent craftsmen, proud of their know-how, run most of the *boulangeries*, where they bake bread in their own ovens. These bakeries stretch across the country, shaping day-to-day social relationships. Industrial production methods struggled to gain a hold in French bakeries, as the work continued primarily as a craft activity, whereas bread was produced industrially in the UK and the US early on, with enormous automated bakeries appearing there in the second half of the nineteenth century. In the UK, the navy's arsenals set up their own industrial bakeries, which played a decisive role in this process. By the end of nineteenth century, John Burnett ([1966] 1989, 122; see also Corley 1976) wrote, "all but the smallest bakehouses had a dough-kneader, and other machinery was being employed for a variety of purposes unimagined a generation earlier".

On the contrary, in France, increasing productivity remained a matter of ongoing controversy (Kaplan 1996; 2002). Even though the first machines for kneading bread appeared as early as the eighteenth century, they only became truly widespread in bakehouses shortly after the First World War. New production demands were not imposed on the basis of an abstract hourly output calculation before the start of the twentieth century. Because of a reduction in manpower as a result of the Great War, as well as the April 23, 1919, legalization of the decreased, eight-hour workday, the issue of bakers' work productivity was fundamentally overhauled. The old concept giving precedence to bodily labor was replaced by output calculations and the triumph of mechanical automata. The discussions surrounding the vote on the law on the eight-hour day show the extent to which modernizing production and equipment was intended to reduce working time by boosting impersonal, mechanized output, compared with the previous representations of productive work based on bodily effort. During the 1920s, the employment minister asked inspectors assigned to investigate the condition of the bakehouses "to what extent improvements in equipment have helped to maintain, over eight hours, the output previously achieved over nine or ten hours, or even to increase this output".[2] This new social norm helped to justify and impose a new representation of mechanized productivity and output, which took more than a century to gain acceptance in the bakery sector (Fridenson and Reynaud 2004, 80).

How can the fact be explained that in France, the organization of work and bread production standards changed so little in the nineteenth century, while the authorities and economists called incessantly for this sector to be industrialized in order to meet demand and resolve the ongoing subsistence problem? In nineteenth-century bakeries, production was highly dependent on physical effort. Bread production was inseparable from the (male) body of the baker, appearing both as an essential tool of the trade and a precondition for making good bread. In fact, the persistence of the craft-based model and the obstacles to industrialization relate primarily to the peculiar status enjoyed by the body as a production tool of this trade. The industrial transformation of bakeries generally went hand in hand with the transformation of the body, its standardization, and even its replacement by a mechanical automaton duplicating and miming the worker's movements. Yet this technology project to remodel the body at work aroused a great deal of resistance on the part of workers, small business owners, and consumers. Thus, throughout the century, the baker's body was at the core of the controversy about defining production standards: It was at the heart of the repeated denunciations of routine, outdated practices, and delays which typified the trade; it was also at the center of the hygiene-related obsession and social fears surrounding workers' suffering bodies. The technological dreams of the modernizers of nineteenth-century bakeries were intended to eliminate the disturbing body of the baker by transferring his skills to a machine.[3]

Thus, the predominant portrayal, which endowed the worker's body with a vital role in bread production, hampered the emergence of the new productivist regime based on an abstract concept of the body, which gained acceptance in Europe and the US between the wars. The "human motor" model and the modern concept of the "workforce" and its machine-based metaphors (see Rabinbach 1992) did not really exist in the nineteenth-century bakery. When exploring the nineteenth-century bakers' world, this, therefore, entails investigating the controversies that surrounded the formalization of the modern, productivist regime based on increases in output and standards of hygiene. The aim is to analyze the change from craft-based working methods founded on the appreciation of the body and its fundamental role in a modern regime with rhetoric and practices based on abstract productivity calculations. Through the example of that highly specific product, bread, this chapter shows how slow and difficult it was for the new productivist industrial regime put forward by contemporary political economy to be adopted in craft-based trades. The particular aim is to understand how the body resisted this evolution and how it was ultimately reclassified in parallel with industrialization and in close conjunction with the controversy regarding views of the body and its status in society.

Bakeries and the Old Regime of Production in the Nineteenth Century

At the beginning of the nineteenth century, bakeries were still rare in rural areas, where the local population had its wheat milled or bought its flour and turned it into dough itself, with the baker often simply taking in the dough for baking. However, during the nineteenth century, the number of bakeries increased everywhere in parallel with urbanization, changes in consumer habits, and a slackening of administrative regulations. They hit a peak number at the start of the twentieth century: 44,000 were recorded in 1901, i.e., one per 874 inhabitants, whereas there was only one per 1,300 inhabitants in 1880. The baker became a familiar yet mysterious figure, both essential and disturbing, especially because of his participation in social unrest. If the bakers were sometimes the targets of riots, especially in the countryside, the relations between masters and journeymen were also a source of trouble because of frequent conflicts among them. In one of the best-known cases, in Lyon, there were 245 bakeries in 1836, 601 in 1876, and 693 in 1906 (Angleraud 1998, 15). In Paris, they were limited in number to 600 until 1863, and then they increased rapidly following the decree of June 22, 1863, which liberalized the creation of bakeries in France: 1,334 were recorded in 1869 and over 2,000 around 1910 (Egrot 1956). In that era, as Roger Thabault ([1943] 1993, 168) noted in his famous sociological work inspired by his village, even in the country "hardly anyone made their own bread anymore . . . they bought big round loaves from the baker". In rural *Départements* such as the Sarthe or Côte-d'Or, most of the bakeries recorded in 1847 were still located in towns—there were around fifty in Le Mans or Dijon, for instance—yet the number of rural bakehouses was growing.[4] In view of this growth, the weak state of the large-scale bakeries puzzled observers (Join-Lembert 1899, 144): How could the repeated failures to produce bread in volume using mechanical methods in cooperative bakeries or in municipal and private businesses be explained?

The history of bakeries and of bread production is not that of the grain trade alone, nor its regulation by the state authorities, but it is a history of bodies at work, a physical history of body movements and techniques. In the nineteenth century, working conditions in small craft-based bakeries had changed little in comparison with those of the eighteenth century, as minutely described by Steven Kaplan (1996). As it appears in the following engraving published by Julien Turgan in the 1860s (Figure 6.1), at a time when the first labor mechanization trials accelerated, producing bread always included three main stages: preparing the yeast, bread making, and baking. First, the yeast had to be made ready, a delicate task that remained the same prior to the introduction of chemical yeast at the end of the nineteenth century. Bread making, which came next, included several stages: the dilution process, which consisted of mixing the yeast and the water; tempering, when flour was added; and, finally, kneading. If the dough was too thick, adding salt and water thinned it. It was then left to rest, well protected

Figure 6.1 Workers in an eighteenth-century bakery (Turgan 1863, 104).

from any draft, which is why old wooden mixing bowls often had lids. Then came the dividing and molding stage followed by a waiting period to allow the dough to rise. Each of these steps involved controlling hand movements precisely in order to ensure the quality of the bread; the expertise of the bakers in making these hand movements was decisive. Kneading in particular was both the hardest and most decisive task for bread quality. Using rapid gestures, the worker had to thicken the dough: He extended his open hands, grasped the dough, carried it, lifted it, folded it, and let it fall in order to introduce as much air as possible. This task usually lasted one and a half hours, demanding know-how and strength and requiring grueling physical effort to do the work (Angleraud 1998, 138; Kaplan 2002, 48–9).

At the beginning of the nineteenth century, physical techniques varied significantly depending on the region, local traditions, and the types of product made by the baker. For instance, in regions that produced bread from stiff dough, the feet were used to knead the dough, which was the case in Spain, Italy, and the Prussian Rhineland. In France, using feet to knead the dough was also carried out in Normandy where hard-crust ("brillé" or "breyé") bread, made from very firm dough, was widespread. In Provence and Marseille, bread was made from flour to which semolina was added, which made working it by hand a very arduous task (Morel 1906). At the end of the eighteenth century, when the Geneva authorities went so far as to invite two baker's boys from Marseille to present their craft to the city's bakers, to widespread astonishment, it was discovered that they used their feet to knead the dough (Wiedmer 1993, 233). These highly diverse physical techniques were difficult to standardize using the production-oriented rules published in professional technical manuals and promoted during trade fairs. Depending on the type of raw ingredients and products to be made, working practices and physical effort would vary considerably.

For bakers, the body was long considered the primary tool of the trade and vital for proper control of the production process. The complex alchemy of bread making was achieved thanks to the body. Although there are few sources for these proletarian portrayals of the body for which no theories were developed, they can be glimpsed through the rhetoric used by technicians, social observers, and doctors. For many of those involved in the bakery sector, mechanization threatened bread quality, because only knowledge possessed by the body made bread production possible. The body and its techniques continued to be perceived as essential by bread makers because they were both the tool and the "ingredient" required to make good bread. The body was a tool because it enabled temperatures to be measured and dough to be warmed. The chemist Henri-François Gaulthier de Claubry, author of the 1838 report on mechanical kneading, thus referred to "the kneader's habit of touching the dough to judge its temperature" and explained:

> A widespread opinion among many bakers is worth investigating . . . they claim that the body heat of the worker kneading the dough warms it up, facilitating proper fermentation, and several bakers have gone so far as to claim that they could tell the difference between bread made by hand and bread made mechanically, by the nature of the mass which had risen far better in the former case, owing to the warmth that it had received for that reason during kneading.
>
> (Gaulthier de Claubry 1839, 19)

Thus, the body itself was involved in production: By imparting its warmth to the dough, it made the fermentation process possible. Other authors emphasized the extent to which direct contact with the body enabled the worker to control the production process. Thus, the authors of the Roret manual point out that

> the worker's arm and hand are both a thermometer and a tool capable of measuring the density and malleability of the dough; it is with the help of these tools that he can get to the exact point where his dough has the consistency, softness, and all the desirable qualities of well-kneaded dough.
>
> (Benoit et al. 1846, 86)

The hand judged both the temperature of the dough and its ability to stretch without breaking. Conversely, "in kneading machines, everything is subject to chance or time, with the worker losing this ability to judge the quality of the bread by touch that is only provided by handling and daily manual work". The use of machines led to a loss of control of the production process, subjecting bread making to the randomness of a machine without "feelings".

Under the Second French Empire, the baker A. Boland (1854) found this analysis deplorable and regretted that some bakers were still convinced

> that men's arms imparted to the dough a warmth that metal, on the contrary, was bound to withdraw. Others, who could be termed ignorant, convinced by the opinion of a scientist on whom his widely recognized merit confers a certain popular authority, believed that the occasionally acidic ammoniac secretions released by a man's body through the grueling task of kneading were a contributor to enhancing fermentation.

The doctor and chemist, François-Vincent Raspail (1840, 334–5), had effectively provided legitimate grounds for the opinion of those in the profession, by asserting, "the best mechanism was still the perfect baker's assistant who grabs the dough with both hands and throws it with all his force into the bottom of the bowl". For Raspail, the sweat generated by the worker was a key component of the production process, making the use of mechanical devices undesirable. In his *Nouveau système de chimie organique*, he explained that the sweat produced by the baker's assistant "adds something to the gluten which increases its elasticity or flexibility" (1840, 335). Thus, he added the endorsement of his scientific prestige to the bakers' arguments against the new machines, which "add nothing similar" to the dough. Still in 1868, the mechanical engineer Louis Lebaudy (1868, 19), despite being in favor of these new processes, asserted that

> the workers, by kneading with their arms and hands, feel the fermentation taking place and, by instinct and practice, can accurately judge the effort required based on the rules used for kneading. A metal or wooden device does not possess this intelligence.

This defense of the superiority of manual labor was rooted an in analysis of the productive uses of the body and the ongoing renewal of the secrets of the craft that permeated the bakers' culture. For Boland (1860, 361), however, there was no doubt that "these revolting suppositions" which "are not confirmed by any theory nor any seriously studied fact . . . have greatly contributed to hampering advances in bread making".

Threatening Bodies, Unproductive Bodies

Bakers and consumers generally shared the conviction that kneading by hand was essential to achieve the taste, color, crispness, and nutritional quality of good bread. At the end of the nineteenth century, social economy theorists such as Charles Gide expressed regret that "consumers [are opposed] to use of the kneading machine, on the grounds that the bread might not be so good".[5] Even at the start of the twentieth century, despite the increased pace of mechanization, many bakers still refused to buy kneading machines because their customers had "an indisputable prejudice against them"; moreover, they asserted, "the doctors are wrong, bread kneaded by hand is far cleaner than mechanically-made bread, the latter being dirtied by the machine itself" (Le Social 1910). In addition, in *Le Petit Parisien*, "the newspaper with the widest circulation in the whole world", as proclaimed on its front page, an article published in 1910 refers to the fact that the use of kneading machines remained very restricted at that time. In order to explain this rejection of machines, the author mentions the prejudice that "bread made by machines was of lower quality", the owners' mistrust of "expensive, cumbersome, and noisy motors", and, lastly, the hostility of workers who fear being "replaced by machines" (Frollo 1910, 1).

It took the massive rise in hygiene-related concerns to discredit these arguments and to convince workers and consumers of the benefits of machine-made bread. The industrial transformation of bakeries involved the prior redefinition of bodily labor, with the reclassification of the baker's body as a threat and a hindrance to advances in both productivity and hygiene. The baker's work was indeed time consuming and grueling, all the more so for it being performed during the night in the poorly aired confines of the bakehouse. Moreover, the division of labor remained sketchy in nineteenth-century bakeries. In general, the baker's only help was a single worker responsible for kneading and many other tasks.[6] These small bakeries could not afford the new machines operated by steam engine. Only at the beginning of the twentieth century, as with many other activities, did the electric motor make it possible and easier to drive modern and flexible machines.[7]

Before that time, only the largest bakeries and cooperatives gradually distinguished between several functions and endeavored to rationalize the work: In these large organizations, a foreman, who was responsible for preparing the dough and loading it into the oven, i.e., the least arduous and most technical tasks, led the team of bakers' assistants tasked with kneading. In craft bakehouses, however, the baker's boy in charge of kneading worked all night from 8:00 p.m. until dawn. Thus, he was considered to be subject to various illnesses and permanently exhausted. Ramazzini ([1777] 1855, 95) asserted, "bakers suffer more illness than other workers". Furthermore, kneading left lasting marks on the body: Bakers' hands, for instance, were "of enormous size . . . you need only see their hands to guess their trade". Likewise, inventories made after death, literary accounts, and hygienists' reports repeatedly stressed how hard this work was. The bakehouse was often only an overheated cubbyhole conducive to all types of illnesses.

Very early on, images of the work recorded these difficulties and descriptions of the craft were filled with endless portrayals of bodily suffering. Folklorists such as Paul Sébillot also referred to the laments of the "bakers" themselves: "A body such as mine is not made of metal, too delicate it is for such a harsh hell", exclaimed a worker.[8] In the nineteenth century, descriptions of the baker's suffering body proliferated and were widely distributed. In 1883, in a "vision of hell", caricaturist André Gill (1883, 234–5) portrayed "half-naked men, reddened by the fire of the ovens, their backs bent and twisted in the night-time wind, raising thick, heavy dough in their muscular arms". In his *Croquis parisiens*, published in the same era, the writer Joris-Karl Huysmans described how "the bodies ran with sweat, the rounded biceps danced on their arms, big beads of sweat formed on their foreheads and soaked up the flour gathered on their temples. They beat the pile like fury" (Huysmans 1905, 68). In a contemporary inquiry, the body of the baker's assistant is compared to that of animals or of "cannibals" rapt in strange trances:

> O weary Pierrots, baker's assistants! . . . sweat, moan and puff; begin your war song and your cannibal dances around the screaming dough! Feast like wolves and drink like the sands, you share your surging chants with the God of the poor: give us our daily bread, o white wrestlers.
>
> (Godart 1910, 4–5)

In view of the bakers' rhetoric in defense of the body's usefulness, hygienists and bakery modernizers attempted to promote a new tamed and "docile" body (Braun 1991). Industrial health awareness, faithful to the Ramazzini tradition, paid great attention to the working area, its ventilation, lighting, and heating. Increasingly, in this literature, the baker's body at work crystallized social fears and became a threat to public health (Bourdelais 2001; Moriceau 2009). Thus, the debate on the mechanization of kneading is inseparable from the increase in sanitary requirements over the nineteenth century. The contact between the worker's skin, covered in large drops of sweat, and the dough seemed in growing contradiction with the basic rules of hygiene. As early as 1829, scientists asserted that, thanks to the kneading machine, "kneading is carried out cleanly, without the need to touch the dough. It is desirable that its use spreads in the interest of cleanliness" (Journal de chimie médicale 1829, 383). In 1838, the prefect-appointed commission for testing kneading machines noted,

> it is obvious that the dough is less clean than if it had not been in contact with men's bodies and, if it could also be made by any other means, this should be used in preference to manual labor. In kneading by machine, workers can only touch the dough when putting it onto the tray.
>
> (Gaulthier de Claubry 1839, 36)

Thus, mechanization was justified in terms of distancing the dough from the body. The main benefit of mechanical methods was

> to ensure, finally, that the prime foodstuff . . . will no longer be drenched in the sweat of those preparing it; that it will be free from the dirt allowed to enter it by people's negligence, excessive fatigue, and personal state through manual kneading or kneading performed using the feet.
>
> (Benoit et al. 1846, 52)

The stigma attached to the baker's body in the name of science enabled the old practices to be increasingly discredited. One of the main advantages of machines was their capacity to effectively create a barrier between the dough and the worker's sweating body, gradually endowed with all the turn of the century's social fears concerning the control of epidemic diseases such as tuberculosis. For the reformers, the drawbacks of manual work and the unsanitary nature of the body threatened sacred bread itself. The industrialization of bakeries and mechanization of the work—particularly the main kneading stage—was thus conducted in the name of cleanliness and public health. Dr. Rigaud stated this already in 1862:

> If one does not feel some pity for the baker's arduous work, one is at least filled with a deep-seated feeling of disgust when seeing these men, most of whose cleanliness is highly problematical, thrusting into the dough their seldom washed hands, their arms dripping with sweat and from whose bodies nauseating emanations may come.
>
> (7)

In his history of the professions for young people, André Parmentier (1908, 5) also asserted, "bread prepared by machines is healthier, cleaner, and more appetizing".

In the second half of the nineteenth century, descriptions of the baker's body afflicted by illness increased in number. Whereas public opinion was stirred by the lack of hygiene in manual work, some went so far as to claim that workers with infectious diseases might contaminate the bread. During the international tuberculosis conference that was held in Paris in 1905, the report by Dr. Jacques Barral (1906, 850) raised the specter of bread causing rapid contagion. In 1906, moreover, it was reported that the municipal council of the town of Elberfeld in Germany had issued a decree prohibiting the use of feet to knead bread, as "this method, in the opinion of the doctors consulted to this end, potentially promoting the spread of infectious diseases" (Revue du Service de l'intendance militaire 1906, 98). In 1910, during a conference organized by the Purchasers' Social League, Professor Letulle confirmed that working conditions in bakeries promoted tuberculosis (Barnes 1995; Angleraud 1998, 150). The bakehouse became a place where the fears of the time crystallized: This was where contagious diseases spread into society. Bread, a sacred, vital food, was transformed into an evil product.

Machines alone seemed capable of halting the danger by replacing the organic body of the baker with the machine's metal organs (Perboyre 1912, 49). At the end of the nineteenth century, doctors and consumers, organized into groups such as the aforementioned Purchasers' Social League, increasingly took action in favor of mechanization; it seemed now the best way to improve the nutritional quality of bread, while cutting the amount of energy needed to make it. They sent petitions to town councils to demand the general introduction of the new mechanical methods (Chessel 2003).

The machine manufacturers also utilized these concerns in their promotion campaigns, and the stigmatization of the baker's body gradually became an advertising ploy. In 1906, a photographic report on an entirely mechanized factory located in La Villette (Paris) emphasized the fact that "here, it is no longer the baker's assistant covered in sweat that does the mixing: it is a machine with muscles of steel" (La Vie illustrée 1906, 221). The metaphor of the body now became ubiquitous when describing machines, which had to perform the function of kneading the dough more efficiently while avoiding the impurities and harmful substances associated with the human body. Whereas the craft baker's bakehouse became a "source of tuberculosis", the new rationalized workshops filled with machines directed by intelligent workers were supposed to produce "perfectly clean" dough.

At the same time, the concept of productivity as applied to bakeries began to spread, particularly after 1860, providing legitimate grounds for mechanization. In popular scientific publications, the notion of productivity first appeared in order to justify the use of machines. In the *Musée de l'industrie*, published in Belgium and circulated widely in France, an 1865 article noted,

> that, while in all other branches of industry, manual labor has, as far as possible, been sidelined by machines, these days, bakeries still work with the same mechanical tools and in exactly the same way as in a bygone era.

This peculiarity may have been on account of the difference between the confined space in the bakehouses and the large size of the machines available. To overcome this obstacle, the spread of small-scale technology was needed, which "by its dimensions, its productivity and its price is suited to all master-bakers in large and small towns" (Bulletin du Musée de l'industrie 1865, 186). From the 1860s onward, at the very time when a new science of work rationalization was emerging, the worker's body was reconsidered from a new perspective: people increasingly sought objective measurements of its output, its strength, the obstacles that hampered it, and the means of improving productivity through the increasing use of machines that replaced men's arms and hands.

The Bread-Making Machine

As early as the eighteenth century, machines appeared to be the solution to ease physical suffering and avert the threats associated with the previous working methods. According to an instructive, possibly apocryphal story often told in nineteenth-century writings, the philosopher Denis Diderot himself may have been moved at the sight of baker's assistants kneading dough. On observing a bakehouse, he apparently provoked the hostility of the workers by asserting, "that such work, instead of being carried out by men's hands, spending their strength, should be done using a machine" (Vinçard 1863, 35). With the advent of the modern era, endeavors to mechanize kneading increased; in Italy, in particular, at the end of the eighteenth century, Genoa's municipal bakeries used kneading machines based on a simple rotating mechanism: A wheel drove vertical bars located in the bowl. In France, the first kneading machines were perfected at the end of the eighteenth century by a baker called Lembert who won an award from the Société d'encouragement pour l'industrie nationale in 1811 (Giedion 1948, 170).

Use of kneading machines, however, was slow to spread, with the number of models growing after 1830, when patents and experiments increased.[9] In 1830, the Hospice de Sedan and the College of Epernay in the Marne region were using a kneading machine, which enabled them to "manage the time and reduce the hardship of the people working in the bakery" (see Bulletin de la Société 1830). In the 1840s, the Mouchot brothers' enormous bakery in Montrouge (Île-de-France) was the first in France to operate on an industrial basis (see Figures 6.2 and 6.3). Thanks to their kneading machines, the jury of the 1844 industrial exhibition in Paris wrote, "the unsanitary and noisy work of the baker's assistants has been eliminated . . . kneading the dough now made more thoroughly and cleaner [has been] freed from men's negligence, coalitions, and illnesses"[10].

Yet the workers and small producers were frequently opposed to this mechanization of the bakehouses and challenged the new rhetoric based on machine-related productivity that was promoted by the makers of the machines and hygienists. In their eyes, in fact, machines introduced new constraints in how work was organized: They were difficult and time consuming to clean, which reduced the time gained over manual kneading. They also introduced additional constraints compared with manual labor that limited the return on investment: machines required more yeast, for example, thus giving the bread a flavor "which is not to everybody's taste" (Benoit et al. 1846, 85–6). They were difficult to convert into hand-operated kneading machines, which meant stopping production if a spare part was not available. In fact, the lack of interchangeability of these first machines' components often hampered their spreading to small bakeries. Moreover, kneading machines disrupted the means of transferring knowledge of the craft: They required a type of training that was ill suited to the traditional system of knowledge transfer handed down from the *Ancien regime* (Angleraud 1998, 74). Finally, the difficulties of mechanization were also connected with the uncertainties surrounding work productivity: In the first half of the nineteenth century, the body was still considered to be a preferable and more profitable tool than expensive machines that were undoubtedly ingenious in theory but unusable in practice (Samuel 1992).

The industrialization of bakeries accelerated after 1860. From this date, the liberalization of the international grain trade, the growth of the railways, the freedom of establishment, and the reduced fears of famine led to an in-depth transformation of the bakery sector. It was increasingly divided

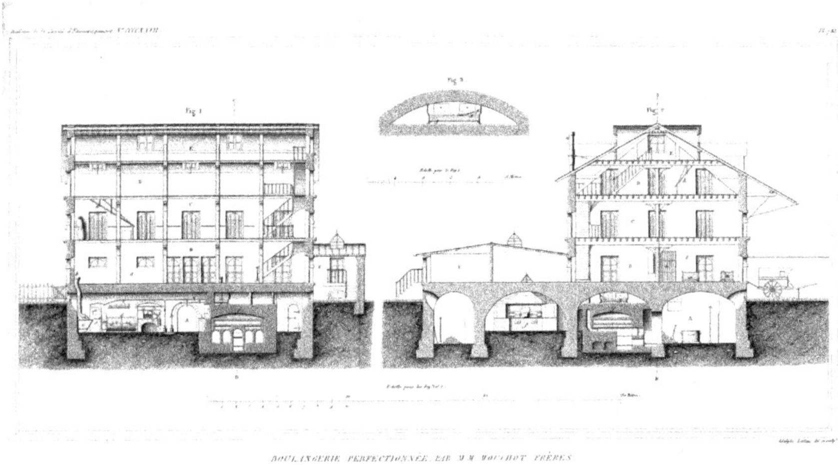

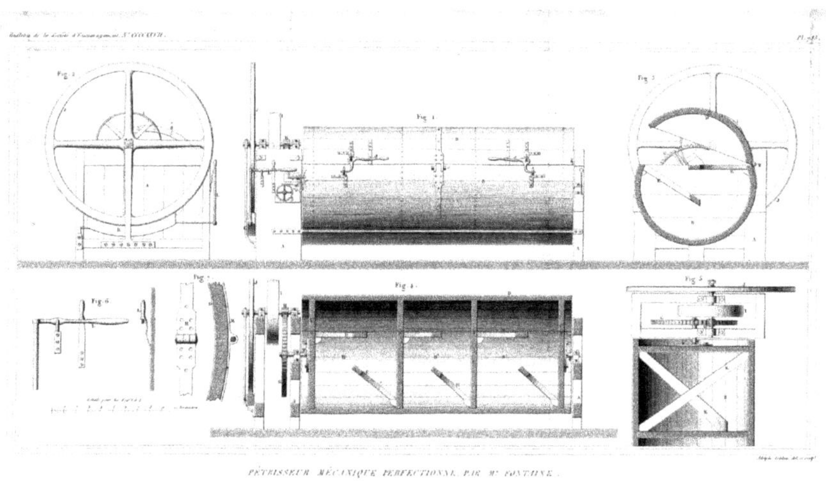

Figures 6.2 and 6.3 The Mouchot brothers' improved bakery, and the Fontaine kneading machine in 1840 (Bulletin de la Société 1840).

into large businesses seeking profit using mechanical resources and the network of small craft bakeries that were still dependent on the worker's body for production. The history of the industrialization of bread making is still not widely known, but the new science of work rationalization and the obsession with productivity clearly emerged in the large bakeries, those established by municipalities and organizations such as the army or schools, as well as in cooperatives (see, for example, Rocquigny 1899). In order to reduce the price of bread, philanthropists and social reformers took a keen interest in the productivity issue too. Anxious to provide the population with not only healthy but also cheap bread, they campaigned for a rationalization of the work, improving production, thanks to a reorganization of the work processes and a redefinition of the worker's body. As such knowledge was widely circulating around the world, these reformers were able to study the diverse models of machines attentively.[11]

Throughout this period, the fight raged on between modernizing industrialists, eager to rationalize work in order to increase productivity, and the mass of small bakers, anxious to preserve their arts and crafts. In Rennes in 1864, for instance, the "Manutention civile" was established, which rapidly became a large factory combining a flour mill and bakery services with new marketing and production methods, including a steam engine to drive the kneading machines. The founders invested the significant sum of 300,000 Francs in equipment; by 1868, the factory was producing seven to eight thousand kilograms of bread per day and supplying approximately fifteen thousand people in the town and its surrounding area (Le Brun 1987). This type of establishment became a focal point of debate: On the one hand, the report by the jury of the 1867 World Fair noted, "bakeries have, in their turn, been invaded by improved processes borrowed from the arsenal of mechanization". Describing the various competing models, the author of the report argued that the kneading machine would enable "the elimination of the work of the baker's assistant, which is so unattractive for the consumer and so hard for the worker, whose naked body runs with sweat" (Rapports du jury international 1868, 118).

On the other hand, small business owners and bakers' boys were opposed to these degrading portrayals of the worker's body, yet their arguments seem to have increasingly become unacceptable. The former feared the scale of the investments that mechanization would demand, whereas the latter were afraid of unemployment. In 1867, the bakers' delegates at the World Fair were aware of the increased pace of mechanization and the growing rejection of manual labor. Instead of resisting the new processes, they endeavored to define what might be a "moral" use of machines—that is, a use that was reconcilable with the interests of the workforce:

> The mistake made by the first bakers that wished to use kneading machines was to imagine that this would allow them to do without bakery workers, or to believe that they could significantly reduce their staff. That was not where they needed to seek to benefit but in economizing on raw materials and in achieving better and higher output.
>
> (Exposition universelle de 1867; 1869)

According to their opinion, mechanization could be of benefit to workers by improving shop floor hygiene and alleviating arduous tasks. For the workers, machines should not replace bodily labor; instead of eliminating the body, they should become a condition for its development and protection. In lieu of relegating the bakery worker to the rank of a docile assistant to the machines, kneading machines could, on the contrary, enable the redeployment of workers:

> Kneading machines require intelligent, competent workers. Mechanization is a brute force that knowledge of the craft and intelligence should direct and control. Workers are obliged to prepare their first yeast, which alone should contain sufficient strength for proper fermentation of the dough, by hand.

After 1890, the number of mechanized bakeries grew further, in particular with the advent of electricity, which removed the obstacles previously created by the use of steam-driven machines. This new source of energy, which was cleaner and better suited to the confined space of the bakehouse, undermined worker opposition, about which the leaders of the cooperative movement complained constantly, while the number of cooperative bakeries in France reached the record number of seven hundred around 1900. The productivity of dough-kneading machines lay at the heart of these shop-floor conflicts, with workers deliberately sabotaging the machines in order to reduce output. "In several cooperative societies, overcoming the workers' resistance and ill will has not yet been achieved", Henri Rivet (1904, 25) noted at that time. At the La Lorraine cooperative in Dombasle-sur-Meurthe, the Déliry kneading machine had to be abandoned after three years of operation because the workers systematically "spoiling" the work convinced the owners "that machine-made bread was less good than bread made by hand". At the Coopération Socialiste in rue Barrault in Paris, at the La Ruche Berruyère in Bourges, or the Cooperative Bakery in Saumur, mechanical automata had to be abandoned "through the fault of the workers who did not know how to use them, because they clearly did not wish to take the trouble to do so". At another cooperative where new working methods had just been implemented, the board of directors was astonished to note that, contrary to its expectations, output did not increase but soon "it was discovered that workers discontented with the new type of work to which they were subjected were burning the flour in the oven to prove the kneading machine's shortcomings from an output standpoint" (28).

Despite such setbacks, the government and the municipalities encouraged mechanization and promoted the new productivist methods. From 1885 on, international flour milling and bakery fairs were organized in order to present the latest mechanical methods. The government granted the tax benefits demanded by business owners: In 1912, for instance, Parliament passed an amendment that exempted them from additional license duty if they installed a kneading machine (see La Boulangerie française 1908; Le Moniteur de la boulangerie 1912). The machine manufacturers and largest bakery owners encouraged emulation by organizing competitions and public demonstrations of the machines, by opening share subscriptions, and by filling the business press with advertisements for the latest innovations, which claimed to increase output while protecting public hygiene.[12] British bakers, who "work with their heads as well as their hands", acted as a model for the modernizers because, unlike French bakers, the "saying 'time is money'" was supposedly their main concern; "machine-made bread is their ideal, weighing machines and dividing machines are highly appreciated there" (L'ami de la Boulangerie 1901).

In the extensive popular literature on these new methods, the body remained the focal point of the analysis. The spread of the machines entailed the invention of a new body at work: In the 1880s, the Parisian

Figure 6.4 Kneading machine with automatic tipping by Werner and Pfleiderer (1887).

manufacturer Arbey vaunted its small "Bolland" kneading machine models operated by a frail-bodied girl. The point was to show both that the work was easier and the possibility provided by the machines of using more docile labor (La Meunerie-boulangerie 1885). In the case of other, more advanced models, the old image of the naked, suffering body of the baker's assistant was replaced by the body of an elite worker wearing a suit, as shown by the image of the "Kneading machine with automatic tipping" exhibited in London during Queen Victoria's Jubilee in 1887 and reproduced in Figure 6.4.

The adoption of machines entailed demonstrating their ability to eliminate workers' bodies as well as their capacity to assimilate the previous efforts of the worker's body; descriptions of machines constantly emphasized the machines' ability to imitate the body of the workers. The parts of the machine were compared to steel arms mirroring the baker's gestures: Machines "knead, divide up, flatten, round, firm up, knead again, imitating the worker's movements with their arms of metal and their wheels, their to-ing and fro-ing, swaying and rotating", noted *L'ami de la boulangerie*, for instance, in 1906. Thus, far from making the body disappear, machines were supposed to improve on it by drawing inspiration from its workings. To market them to bakers, advertising constantly extolled the machines' organic workings, their ability to retain the benefits of the arms of the baker's boy while avoiding their dangers. Whereas bakers used to be male workers, some pictures also represented machines driven by women and children, although primarily for advertising purposes, so as to show how it was easy to use the new "Bolland" kneading machine mentioned earlier.

Conclusion

Throughout the nineteenth century, the new productivist regime of industrial capitalism was slow to gain acceptance in French bakeries. The bread issue, the symbolic and political stakes involved and the craft-based production methods used, made it an unusual product that was difficult to standardize using the new industrial norms. However, the situation changed quite rapidly at the start of the twentieth century, first in Paris and in other large cities and then, more slowly, in provincial bakeries. In 1908 Lyon, 250 kneading machines were already working in the city's 750 bakeries. When the war broke out, most bakeries were equipped with machines and work inspectors explained this early uptake in Lyon by "the electrification of the bakehouses being made exceptionally easy in a region where electricity is commonly available thanks to the proximity of hydroelectricity"[13]. The proportion was slightly lower in Paris, as only two hundred bakeries were equipped with kneading machines out of the total two thousand shops that were recorded in 1908. In a rural *Département* such as the Nièvre, bakers were just starting to acquire machines, yet the lack of apprentices finally forced them to replace the body of the baker's boy with mechanical organs

(Thuillier 1965). Yet it was the First World War, in particular, that accelerated the transition to the new regime of industrial productivity.

The massive mobilization of conscripts in 1914 forced many bakers to install machines in their bakehouses in order to make up for the lack of manpower. By 1917, 50 percent of Parisian bakeries were equipped with kneading machines, and the remainder made the leap during the 1920s.[14] Even though the situation varied greatly from one region to another, a consensus emerged concerning the benefits of kneading machines in terms of accelerating production; machines were supposed to eliminate night work and create more "joy of work" and "embellishment of workplaces",[15] even though many owners continued to assert that "with kneading machines, output has not increased, the work is just less difficult for the workers whose presence is nonetheless needed".[16] Thus, an extensive mechanization of bakeries occurred just when an abstract concept of productivity started to be widely debated (see this volume's introduction). In 1911, Albert Aftalion defined productivity as "the ratio of the total production obtained in a given time and by all the staff employed in this production". In the interwar period, productivity was at the heart of the concerns of engineers and economics anxious to implement American methods of rationalization of work in France (Moutet 1997; Boulat 2008). Although economic analysis never centered on bakeries per se, the slogan of increasing productivity triumphed also in this context. Popular science magazines diffused this perspective, and bosses converted to the machines because they hoped that this would replace the problematic human labor force while increasing their profits (see Revue générale des sciences 1931).

At the same time, the experience of French bakeries in the nineteenth century led to a profoundly ambivalent situation. Despite the triumph of the postwar industrial regime, a special aura continued to surround the baker's productive body; the baker, with his skilled manual craftsmanship, remains the yardstick in judging bread quality till today. Thus, the example of French bakeries in the nineteenth century ultimately shows how the emergence of the modern productivist regime was a conflict-ridden, controversial process, one that was slow to gain acceptance through the reclassification of the body at work, the imposition of new mechanical automata, and the promotion of a new knowledge of hygiene and chemistry in the bread-making process. Yet this new productivity regime remained contested: Even nowadays, kneading by hand seems to be reemerging in conjunction with the fashion for organic bread, with the contact of the worker's body with the dough expressing an act of protest against the devastating outcomes of the productivist regime of the agro-food industry.

Notes

1 See, for example, Confédération Nationale de la Boulangerie-Pâtisserie Française: http://www.boulangerie.org/economie/economie-panorama-du-secteur.

2 French National Archives (Paris), F22 567: "Enquête sur le développement du machinisme dans la boulangerie (1924). Circulaire du 19 août 1924".
3 This research continues previous research; see, in particular, François Jarrige (2010).
4 French National Archives (Paris), F11 2771: "Statistiques du nombre de boulangeries en France"; Departmental Archives of Sarthe (Le Mans), 6 M 318: "Nombre de boulangers dans le département et importance de leur production (1847)".
5 Ministère du Travail et de la Prévoyance sociale, Conseil Supérieur du travail, Compte rendu, Paris, Session de 1911: le travail de nuit dans les boulangeries, séance du vendredi 17 novembre 1911, 63.
6 In Lyon, the number of workers by bakery remained stable during the nineteenth century: 0.9 on average in 1836, 1.18 in 1856, and 1.06 in 1896 (Angleraud 1998, 144).
7 Regarding the vitality of small firms and alternative flexible technology in nineteenth century, see Charles Sabel and Jonathan Zeitlin (1985).
8 *La misère des garçons boulangers de la ville et des faubourgs de Paris*, Troyes, 1739, quoted in Paul Sébillot (1894–1895, 19).
9 Marcel Arpin (1948) calculates that nine devices were invented between 1830 and 1840, thirteen between 1840 and 1850, fifty-seven between 1850 and 1860, fifty-two between 1860 and 1870, twenty-one between 1870 and 1880, forty-eight between 1880 and 1890, fifty-four between 1890 and 1900, and the number increases further after that date.
10 They received a gold medal at the 1844 Industrial Exhibition in Paris (see Exposition des produits de l'industrie française 1844, 790).
11 If foreign machinery are imported and widely described in France, the press of the US also describes experiments in Paris (see e.g. Scientific American 1885).
12 Regarding these strategies to convince of the utility of mechanical kneading, see François Jarrige (2010, 671–5).
13 French National Archives (Paris), F22 567: "Enquête sur le développement du machinisme dans la boulangerie (1924), Lyon, le 13 mars 1924, rapport de l'inspecteur divisionnaire du travail à Lyon".
14 French National Archives (Paris) F22 567: "Enquête sur le développement du machinisme dans la boulangerie" (1924); and Raymond Calvel (1952).
15 On the discourses of "joy of work" that emerged in the first decades of the twentieth century, see the book by Joan Campbell (1989).
16 French National Archives (Paris), F22 567: rapport de M. Py, Caen, 10 Septembre 1924.

Bibliography

Angleraud, Bernadette. 1998. *Les boulangers Lyonnais aux XIX-XXe siècles*. Paris: Éditions Christian.

Arpin, Marcel. 1948. *Historique de la meunerie et de la boulangerie*. Paris: Le Chancelier.

Barnes, David S. 1995. *The Making of a Social Disease: Tuberculosis in Nineteenth-Century France*. Berkeley: University of California Press.

Barral, Jacques. 1906. "La panification et la tuberculose." In *Congrès international de la tuberculose, tenu à Paris, du 2 au 7 octobre 1905*. Paris: Masson.

Benoit, Philippe-Martin-Narcisse, Julia de Fontenelle, and Jean-Sébastien-Eugène Malepeyre. 1846. *Nouveau manuel complet du boulanger, du négociant en grain, du meunier et du constructeur de moulins*. Vol. 2. Paris: Roret.

Boland, A. 1854. *Observations sur l'application de la mécanique à la boulangerie. Nouveau pétrisseur mécanique inventé par Boland*. Paris: Mme Bouchard-Huzard.
———. 1860. *Traité pratique de boulangerie*. Paris: Lacroix.
Boulat, Régis. 2008. *Jean Fourastié, un expert en productivité et en modernisation*. Besançon: PUFC.
Bourdelais, Patrice, ed. 2001. *Les Hygiénistes: enjeux, modèles et pratiques*. Paris: Éditions Belin.
Braun, Rudolf. 1991. "The 'Docile' Body as an Economic-Industrial Growth Factor." In *Favorites of Fortune: Technology, Growth, and Economic Development Since the Industrial Revolution*, edited by Patrice Higonnet, David S. Landes and Henry Rosovsky, 120–41. Cambridge, MA: Harvard University Press.
Bulletin de la Société d'Encouragement pour l'Industrie Nationale. 1830. "Rapport fait par M. Mallet, au nom d'une commission spéciale, sur le pétrisseur mécanique, présenté à la société par MM. Cavalier, frère et cie." 314: 307–14.
———. 1840. "Description de la boulangerie perfectionnée de Mouchot Frères." 427: 785.
Bulletin du Musée de l'industrie. 1865. "Nouvelle machine à pétrir le pain." 47 (April): 186–7.
Burnett, John. (1966) 1989. *Plenty and Want: A Social History of Food in England from 1815 to the Present Day*. London: Routledge.
Calvel, Raymond. 1952. *La boulangerie moderne*. Paris: Eyrolles.
Campbell, Joan. 1989. *Joy in Work, German Work. The National Debate, 1800–1945*. Princeton: Princeton University Press.
Chessel, Marie-Emmanuelle. 2003. "Aux origines de la consommation engagée: la Ligue sociale d'acheteurs (1902–1914)." *Vingtième Siècle* 77: 95–108.
Corley, T.A.B. 1976. "Nutrition, Technology and the Growth of the British Biscuit Industry, 1820–1900." In *The Making of the Modern British Diet*, edited by D.J. Oddy and D. Miller, 13–25. London: Croom Helm.
Egrot, Mlle. 1956. "La boulangerie parisienne sous le second Empire." *L'actualité de l'histoire* 14: 12–29.
Exposition des produits de l'industrie française en 1844: Rapport du jury central. 1844. Vol. 2. Paris: Fain et Thunot.
Exposition universelle de 1867: Rapports des délégations ouvrières. 1869. Vol. 1. Paris: Morel.
Fridenson, Patrick, and Bénédicte Reynaud, eds. 2004. *La France et le temps de travail, 1814–2004*. Paris: Odile Jacob.
Frollo, Jean. 1910. "Dans les boulangeries." *Le Petit parisien*. February 1: 1.
Gaulthier de Claubry, Henri-François. 1839. "Rapport sur la fabrication du pain par le pétrissage à bras et par les machines; fait au nom d'une commission spéciale par M. Gaultier de Claubry." *Annales d'hygiène publique et de médecine légale* XXI: 6–88.
Giedion, Siegfried. 1948. *Mechanization Takes Command: A Contribution to Anonymous History*. Oxford: Oxford University Press.
Gill, André. 1883. *Vingt années de Paris*. Paris: C. Marpon et E. Flammarion.
Godart, Justin. 1910. *Les mineurs blancs*. Paris: La publication sociale.
Huysmans, Joris-Karl. 1905. *Croquis parisiens*. Paris: Stock.
Jarrige, François. 2010. "Le travail de la routine: autour d'une controverse sociotechnique dans la boulangerie française du XIXe siècle." *Annales. Histoire, Sciences Sociales* 65: 645–77.

Join-Lembert, André. 1829. *L'organisation de la boulangerie en France*. Paris: A. Rousseau, 1899.

Journal de chimie médicale. 1829. 7: 383.

Kaplan, Steven L. 1996. *The Bakers of Paris and the Bread Question, 1700–1775*. Durham: Duke University Press.

———. 2002. *Le retour du bon pain. Une histoire contemporaine du pain, de ses techniques et de ses hommes*. Paris: Perrin.

L'ami de la Boulangerie. 1901. "Relation de M. Périer sur l'exposition de la boulangerie à Londres Samedi 6 octobre 1901." (October 6).

———. 1906. "Exposition de boulangerie à Lausanne." (August 30).

La Boulangerie française. 1908. "La boulangerie, les pétrins mécaniques et le fisc." (August 8).

La Meunerie-boulangerie, journal officiel de l'Exposition international. 1885. "Pétrins Boland, à bras et à moteur." 3 (May 20): 46.

La Vie illustrée. 1906. "Les industries parisiennes. L'industrie du pain." 377 (January 4): 220–23.

Lebaudy, Louis. 1868. *Le Pétrissage mécanique en boulangerie*. Paris: Paul Dupont.

Le Brun, Yvonne. 1987. "La liberté de la boulangerie à Rennes sous le Second Empire." *Annales de Bretagne et des pays de l'ouest* 94 (3): 167–84.

Le Moniteur de la boulangerie, organe du syndicat des patrons boulangers de Marseille. 1912. "L'impôt des pétrins mécaniques." (March 15).

Le Social. 1910. "Nos enquêtes. La diffusion du pétrin mécanique." (April 10).

Morel, A. 1906. "Le pétrissage de la pâte avec les pieds." *Bulletin des Halles* (July 31).

Moriceau, Caroline. 2009. *Les Douleurs de l'industrie. L'hygiénisme en France, 1860–1914*. Paris: EHESS.

Moutet, Aimée. 1997. *Les Logiques de l'entreprise, la rationalisation de l'industrie française de l'entre-deux-guerres*. Paris: EHESS.

Parmentier, André. 1908. *Les Métiers et leur histoire*. Paris: A. Colin.

Perboyre, René. 1912. " 'Mineurs blancs' " et tuberculose. Danger social du travail de nuit dans la boulangerie". PhD diss., University of Toulouse.

Rabinbach, Anson. 1992. *The Human Motor: Energy, Fatigue, and the Origins of Modernity*. Berkeley/Los Angeles: University of California Press.

Ramazzini. (1777) 1855. *Traité de la maladie des artisans*. Paris: A. Delahays.

Rapports du jury international. Exposition universelle de 1867. 1868. Paris: P. Dupont.

Raspail, François-Vincent. 1840. *Nouveau système de chimie organique*. Paris: Société encyclographique des sciences médicales.

Revue du Service de l'intendance militaire. 1906. 19: 98–9.

Revue générale des sciences pures et appliquées. 1931. "La panification mécanique." 42: 536–8.

Rigaud, François-Charles. 1862. *Sur la boulangerie du point de vue de l'hygiène public*. Paris: Masson.

Rivet, Henri. 1904. *Les boulangeries coopératives en France*. Paris: Larose.

Rocquigny, Robert de. 1899. "Une enquête sur les boulangeries coopératives rurales." *Bulletin du Musée Social* 12: 565–605.

Sabel, Charles, and Jonathan Zeitlin. 1985. "Historical Alternatives to Mass Production: Politics, Markets and Technology in Nineteenth-Century Industrialization." *Past & Present* 108: 133–76.

Samuel, Raphael. 1992. "Mechanization and Hand Labor in Industrializing Britain." In *The Industrial Revolution and Work in Nineteenth-Century Europe*, edited by Lenard R. Berlanstein, 26–43. London: Routledge.

Scientific American. 1885. "Machines for Bakeries." (October 17).

Sébillot, Paul. 1894–1895. *Légendes et curiosités des métiers*. 20 vols. Paris: Flammarion.

Thabault, Roger. (1943) 1993. *Mon village. Ses hommes, ses routes, son école*. Paris: PFNSP.

Thuillier, Guy. 1965. "Au XIXe siècle: l'alimentation en Nivernais." *Annales. Économies, Sociétés, Civilisations* 20: 1163–84.

Turgan, Julien. 1863. *Les grandes usines de France: tableau de l'industrie française au XIXe siècle*. Vol. 3. Paris: Michel Lévy frères.

Vinçard, Pierre. 1863. "Le garçon boulanger." In *Les ouvriers de Paris: Alimentation*, edited by Pierre Vinçard, 1–69. Paris: Gosselin.

Werner and Pfleiderer. 1887. "Universal Kneeding & Mixing Machine." *Le blé* 35: 179.

Wiedmer, Laurence. 1993. *Pain quotidien et pain de disette. Meuniers, boulangers et Etat nourricier à Genève, XVII–XVIIIe siècles*. Genève: Éditions Passé-présent.

Part II

Transformations of Twentieth-Century Productivism

Introduction to Part II

Peter-Paul Bänziger and Mischa Suter

As the chapters in Part I indicated, during the eighteenth and nineteenth centuries, productivist thinking and the concept of productivity had to battle for predominance against other ways of considering production and economy. Chief among these were the bourgeois discourse of moderation and the communitarian notion of duty propagated by, for example, the Christian churches (see Maß, this volume). Not until the decades around 1900 were they supplanted by an ethos of performance and efficiency (Sarasin 2001; Verheyen 2014) that conceived of production as an abstract, optimizable relationship between input, output, and time. In contrast to the family-structured production communities of the bourgeois era, this form of production was widely detached from the domestic and family sphere and increasingly organized in large-scale factories (see Jarrige, this volume). Key to its ascent were processes of scientization, especially the rise of thermodynamic explanatory models in the second half of the nineteenth century (Rabinbach 1992).

It was against this backdrop that economists such as Irving Fisher reflected on productivity. As Nina Mackert shows, their chief concern was no longer the avoidance of waste on the model of the bourgeois thrift ethos (see Maß, this volume); instead, they aimed to optimize the relationship between food input and labor output. This both increased labor productivity in the factory and took the wind out of the sails of workers' calls for better pay—for what counted was no longer the abstract level of pay, but "right" consumption. At the same time, these debates reflect a fundamental shift in thinking within economics. Increasingly, notions of affluence and an orientation on consumption complemented resource scarcity-oriented analyses of production processes. In this sense, the productivity regime of the Progressive Era in the US may be regarded as epitomizing the consumer and work societies of classical modernity in which consumption-oriented offers of identity were framed by a performance-oriented work ethos and vice versa (Bänziger 2015).

A second context for twentieth-century thinking on productivity is the welfare state, investigated by Alexa Geisthövel through the example of the occupational-therapy regime of productivity in interwar Germany.

At this time, in the context of a concept of citizenship that was grounded in work and health, the state of being "productively" active became the goal of therapeutic assemblages as well; ultimately, health was equated with fitness for work. Productivity here was little more than a synonym for work. Thus, the concept lacked the abstract quality that Max Weber had advocated at the start of the century and that had become largely established in other realms (see the introduction to this volume). Simultaneously, thinking on work and productivity was now increasingly embedded to larger degrees in the growing field of macroeconomics and the theories of the later Historical School of Political Economy. These concepts distinguished the "technical" productivity of individual production factors from the perspective of political economy that also underlay theories of the welfare state.

The political economy perspective was also crucial to the growth paradigm of the midcentury (see the introduction to this volume), as pursued by the Latin American developmentalism discussed by Monika Streule. Through the example of large-scale residential building projects in Mexico City, Streule demonstrates that the concept of productivity so dominant in developmentalist thinking lost ground there to discourses on social reform, culturalism, and nation building. The backward-looking, unproductive peasant was contrasted with the young (male) worker, portrayed as modern and urban. Progress on productivity was envisioned less as an increment in individual performance than as an enlargement of the industrial labor force. Here, too, a tendency to equate productivity with industrial labor can be observed. Furthermore, Streule stresses the importance of spatial aspects of productivity regimes, in this case, the space of a metropolis and its house-building projects. The profound shift in funding modalities—from large public housing projects to private loans for single-family row houses—also marks the transition, since the 1970s, from social engineering in a developmentalist regime of productivity to the growing invocation of individual commitment.

Elsewhere as well, the crises of the 1970s seem to have put a considerable damper on the Fordist productivity regimes of the twentieth century's first two-thirds. This certainly does not mean that the idea of productivity had run its course. At least in microeconomics, it remained a core concept, despite substantial shifts in emphasis, as described by Lukas Held. He argues that, first and foremost, the aim of postwar balance sheet analyses was to retrospectively optimize production processes, for example, by investigating the waste of resources. This practice was not abandoned in the 1970s and 1980s, but it was increasingly edged out by efforts to influence the factors of individual willingness to perform, drawing on the psychology of motivation. Unlike during the productivity regime of the Progressive Era, the issue now was now no longer how to deal with the phenomena of consumer society, but how to be prepared for an apparently unpredictable future.

The chapters in Part II highlight three aspects of the history of productivity regimes. First, productivity remained a central concept for both economic theory and production throughout the twentieth century. Second, however, this does not detract from the fact that productivity remained an essentially contested concept, or that it experienced displacements and changing fortunes. Thus, the macroeconomic growth paradigm, with its comparative productivity analyses, had its heyday in the post-World War II period; it came under increasing pressure from the 1970s on. Yet this very example demonstrates that growth and productivity cannot simply be conflated. Productivity may also play an important role in critiques of growth, for example, in terms of more efficient utilization of existing resources—a potential way of thinking that the rubber producers in German Cameroon described by Tristan Oestermann (this volume) still lacked. However, on the microeconomic side, resource-oriented analyses seem to have gone out of fashion since the 1970s. Yet the long history of "human capital" approaches (see Bernet and Gugerli 2011), which resonates in Mackert's and Geisthövel's essays, indicates that the new theories of the 1970s had their own history, making it impossible to write about the last decades of the twentieth century in terms of radical rupture.

Third, what goes for the concept of productivity in economic theory and production practice applies in equal measure to the body: the specific ways in which it was addressed underwent change. The predominant, thermodynamically informed image of the "human motor" gave way to systems-based, cybernetic, and psychological models. The body did not, however, cease to play a central part in conceptualizing and implementing regimes of productivity. In the twentieth century, human beings and their bodies remained the most important factors of production and productivity (see Kaufmann 2012)—in the face of all the fantasies of automation and perhaps precisely because of the countless prophesies of the "end of work" (Rifkin 1995).

Bibliography

Bänziger, Peter-Paul. 2015. "Von der Arbeits- zur Konsumgesellschaft? Kritik eines Leitmotivs der deutschsprachigen Zeitgeschichtsschreibung." *Zeithistorische Forschungen* 12 (1): 11–38.

Bernet, Brigitta, and David Gugerli. 2011. "Sputniks Resonanzen. Der Aufstieg der Humankapitaltheorie im Kalten Krieg—eine Argumentationsskizze." *Historische Anthropologie* 19 (3): 433–46.

Kaufmann, Therese. 2012. "Materiality of Knowledge." Translated by Aileen Derieg. In *transversal* 1. Accessed January 7, 2016. http://eipcp.net/transversal/0112/kaufmann/en.

Rabinbach, Anson. 1992. *The Human Motor: Energy, Fatigue, and the Origins of Modernity*. Berkeley, CA: University of California Press.

Rifkin, Jeremy. 1995. *The End of Work: The Decline of the Global Labor Force and the Dawn of the Post-Market Era*. New York: Putnam Publishing Group.

Sarasin, Philipp. 2001. *Reizbare Maschinen: Eine Geschichte des Körpers 1765–1914*. Frankfurt am Main: Suhrkamp.

Verheyen, Nina. 2014. "Bürgerliches Leistungsethos? Geschichtswissenschaftliche Korrekturen einer irreführenden Formel." In *Erfolg: Konstellationen und Paradoxien einer gesellschaftlichen Leitorientierung*, edited by Denis Hänzi, Hildegard Matthies, and Dagmar Simon, 45–61. Baden-Baden: Nomos.

7 Feeding Productive Bodies

Calories, Nutritional Values, and Ability in the Progressive-Era US

Nina Mackert

Under the headline "What Does 100 Calories Look Like?", the web portal *SparkPeople* features pictures of plates that contain one-hundred-calorie-servings of fruits, vegetables, nuts, protein, and grains (Megan n.d.).[1] With these and other posts, the US-based company SparkPeople, Inc. (n.d.), which runs a range of health portals, seeks to "help . . . everyone learn to eat better . . . for life". Counting calories is obviously a huge part of this endeavor. Numerous websites from different backgrounds encourage people to eat right, to "keep portion distortion in check" (Dairy Council of California n.d.) in order to lose weight and to stay healthy. "You need energy from foods and drinks to fuel your body—for everything from breathing to physical activity", the USDA (n.d.) states but, at the same time, urges people to keep their "calorie limit in mind when deciding what to eat and drink". When individuals approach supermarket shelves, grab an item, and check its caloric content, the calorie has inscribed itself into their routines. It yields knowledge on eating "right", distinct practices of bodies, the materiality of caloric tables as well as meal sizes, and individuals that understand themselves in relation to the calorie.

In short, the calorie is productive.[2] Moreover, its history is immanently interwoven with the history of productive bodies and negotiations of their abilities and improvement. It is hardly a coincidence that one of the first proponents of one-hundred-calorie portions in the US was, at the same time, preoccupied with ways to increase the nation's capital—including the human capital: In the early twentieth century, Yale economist Irving Fisher, known for his works on capital theory, interest, and debt, sought to improve the lives and wealth of Americans by regulating their intake of calories.[3] He worried that "many people eat too much, that is, too many calories" (Fisher and Fisk 1915, 28) and suggested that such unhealthy diets would burden the nation with huge costs: In his 1909 report, *National Vitality*, he estimated the annual loss of "preventable deaths" and illnesses to be about 1.5 billion dollar (Fisher 1909; see also Nordhaus 2005, 367). Understanding humans as service-yielding capital, Fisher saw huge room for its accumulation through healthy eating and living, and he identified food as a key factor for increased working power (see Fisher 1897, 511–37; 1909,

47; Dimand 2005, 393–7; Nordhaus 2005). In addition, because he deemed previous ways of counting calories too complicated, one of Fisher's (1906) suggested steps to facilitate eating right was to measure any food eaten in one-hundred-calorie portions.

Speaking of previous ways of counting calories is somewhat misleading, however. At that time, healthy dieting and especially the calorie were brand-new concepts. They emerged in an era that was shaped by the belief in the improvement of bodily, social, and economic efficiency by modern science (Wiebe 1967; Lears 2009). Fisher is an embodied example of the nexus between political economy and nutritional science in the Progressive-Era US. In 1898, he had suffered from tuberculosis (Dimand 2005, 393) and since then he published numerous articles, books, and bulletins on (dietary) health and its economic importance, telling the story of his overcoming the disease as a tale of catharsis and the experience of sheer endless possibilities: "At length by dint of conscientious application of a dozen or more specific points of hygiene, not only did I succeed in winning back my previous working power, but acquired more than I had ever dreamed of acquiring" (quoted in Nordhaus 2005, 368). In his promise of new possibilities of health and productivity, Fisher's story was much more than simply one of individual cure. By suggesting an increased productivity through consumption, he intervened into the conflict between producerist and consumerist rationales that contemporaries grappled with in the emerging consumer society (Horowitz 1980; Armstrong 1998, 53; Stoff 2001). The mélange of economic theory and nutritional science mediated this tension, as this chapter will argue, by shaping a modern body that was marked by constant improvement of income and life through productive wants and restrained consumption.

This chapter discusses the calorie as emerging within the establishment of this comprehensive norm of having and improving a healthy and productive body and mind—an imperative that Robert McRuer has called "compulsory able-bodiedness".[4] Following McRuer, compulsory able-bodiedness is understood here as a crucial *dispositif* of the Western late capitalist body order—that is, as a network of knowledge, institutions, and practices that shaped, regulated, and organized bodies since the late nineteenth century.[5] This norm of able-bodiedness can be called compulsory, not because aberrations are disciplined, but because the norm creates them. It is the naturalization of capability that produces bodies as disabled, sick, or deficient. Labeling it compulsory allows us to state, as McRuer (2002, 89) points out, that able-bodiedness is established "as the foundational . . . identity" of bodies and "sets itself up as the origin and the ground" (McRuer 2002, 93; quoting Butler 1991, 21). That means not only that to be able-bodied is the norm but also that everybody must strive for this norm. Not doing this, however, risks rendering one unintelligible. The character of compulsion, thus, is rather one of obligation than of coercion. To speak of compulsory able-bodiedness refers less to the possibility of punishing deviance but to the normalization of able-bodiedness and especially to its imperative for cure.[6] It is barely thinkable to *not* aspire to health, for instance.

Thus, the productivity of able-bodiedness for the liberal order lies in its highly subjectifying character as something for which everyone *must* strive. This is key to this chapter's assessment of compulsory able-bodiedness: With its powerful imperative for prevention, cure, and rehabilitation, this requirement demands working on one's self, be it by exercising, eating, or in any other way, so as to attain a healthy and efficient body. From a perspective of Critical Ability Studies, this chapter focuses on the ways in which compulsory able-bodiedness became such a vantage point of subjectivation.[7]

The following sections will demonstrate how compulsory able-bodiedness became a crucial aspect of what might be called the "regime of productivity" of the Progressive-Era US, and how this norm was shaped by a range of heterogeneous elements such as economic theory, nutritional science, specific modes of consumerism, and a distinct reform ethos. The calorie acted as part and vehicle of this regime, marking new understandings of bodies and carving out ways of eating "right". In order to unfold the role of food and eating in the emergence of compulsory able-bodiedness, this chapter starts by analyzing the discursive network of nutritional knowledge and economic thought, its conceptions of the productive body, and the invention of the calorie at the end of the nineteenth century. Next, this chapter will discuss problems and possibilities that economists and nutritionists saw in different ways of eating and measuring food, as well as the changed politics of inclusion and exclusion that these dietary efforts yielded. Embedding food reform into the turn from producerism to consumerism, this chapter then addresses Fisher's concept of human capital before ending with a brief return to the calorie.

Inventing Caloric Efficiency

For the economist Fisher, food was a key factor in influencing the productivity of individuals and the nation. In his study *National Vitality*, he predicted that the emerging nutritional science would "develop rapidly" (Fisher 1909, 86) and prove insightful in the future. The economist had prepared this report for the National Conservation Commission, founded in 1908 by US president Theodore Roosevelt, so as to encourage the conservation of national "natural resources" such as water, forests, and land. The Progressive conservation campaign was part of a post-frontier idealization of wilderness and outdoor activities. It reacted to the fear that after the closing of the frontier, the absence of physical challenges would soften and effeminate American citizens. Corresponding with imperial fantasies and the fear of racial decline, white male bodies seemed to be endangered by the civilization they were thought to advance (Bederman 1995, sec. 5; Simon 2003). Those allegedly endangered bodies became the centerpiece of Fisher's intervention. "The problem of conserving natural resources is only one part of the larger problem of conserving national efficiency", Fisher (1909, 1) stated at the beginning of his treatise. For him, the improvement of the nation's "vital assets" was even more important, as it promised to "increase . . . economic productivity". Hence, striving for conservation went beyond avoiding waste and conserving resources. Very fittingly, it was, at the same time, a *Progressive* endeavor, in the sense of a biopolitical optimization of bodies, populations, and productivity. "The world is gradually awakening to the fact of its own improvability" (14), Fisher rejoiced in his report and, thereby, perfectly characterized the zeitgeist of the Progressive-Era US: the strong belief in the power of scientific expertise to understand and improve bodies and society (see Wiebe 1967; Lears 2009). Efficiency then denounced the increase of productivity while avoiding waste—this pervasive ideal of the Progressive Era became a model not only for industrial processes but also for bodies (Tichi 1987, sec. 2).

The emergence of modern physiology illustrates that the "scientification" of bodies was driven by efforts to increase their productivity (Sarasin and Tanner 1999, 29; Lears 2009). Numerous experiments and studies by physicists, physicians, chemists, and nutritionists tried to measure and maximize the work, power, and energy of bodies. Together, they constructed a new understanding of bodies that regarded them as engines, burning fuel and transforming it into energy (Sarasin and Tanner 1999, 19). This concept of the "human motor" was based on thermodynamic thought that had been pervading the physical sciences since the middle of the nineteenth century (Tichi 1987; Rabinbach 1990). The first law of thermodynamics, the law of energy conservation, states that the amount of energy in a closed system remains the same, that it can change its form, but it is not lost and cannot be magnified (Mudry 2009, 33; Neswald 2013, 88). Applying this law to human bodies meant to suggest that bodily processes of food transformation did not change the relation of energetic input and output. Thus, because scientists understood bodies as driven by transforming an input of food energy into an output of physical energy—of work—they became occupied with the chemical compositions of foodstuffs and possible ways of optimizing this process (Tanner 1999, 74; Mudry 2009 30–3; Neswald 2013, 89–90). The idea that bodily productivity had to do with its food intake acquired a physiological fundament. The history of thermodynamics is, as Elizabeth Neswald (2013, 103) argues, central to understanding the emergence and course of nutritional science, as it brought an important new measurement unit that allowed explicit quantification of the relation of food and productivity: the calorie.

Whereas it had been European physiologists who had initially measured food in calories, the chemist Wilbur O. Atwater operationalized the calorie for food reform in the US. For a few years in the 1880s, Atwater had been working in the laboratories of the German physiologists Carl von Voit and Max Rubner. Both scientists experimented with calorimeters—devices that allowed the measurement of heat energy—and especially the latter had adapted this research not only to food but also to the human body (Treitel 2008; Mudry 2009, 33–4). In the later 1890s, after a transatlantic transfer of knowledge and devices, Atwater and his team completed the construction of a human calorimeter in his laboratory at Wesleyan University in Middletown, Connecticut. Their test subjects were asked to stay in the apparatus for days and nights, to eat distinct quantities of food, and to alternately exercise and rest. By analyzing their production of heat, carbon dioxide, and feces, the scientists determined the metabolized heat energy, measuring it in calories.[8]

Not accidentally, devices similar to this calorimeter had been used before to measure the efficiency of steam engines (Hargrove 2006, 29–58). That it was now applied to human motors shows that the emergence of the calorie was both the effect and means of establishing a thermodynamic model of bodies. As Nick Cullather (2007, 340–7) has stressed, the productivity of the calorie for the biopolitical order—in the US as well as on the global scale—can hardly be underestimated. By providing an allegedly objective, scientific view on food, the calorie enabled the precise comparison of the diets of different classes and "races". It shaped the nutritional management of factories, prisons, and schools, as well as armies, and it regulated people's diets at home (Mudry 2009, 44; Veit 2013, 45). The calorie wasn't simply a new unit to measure food, but the very foundation of apprehending the efficiency of eating.

The Problem of Protein

Irving Fisher (1906, 417) was a fan of the calorie, too, and he praised Atwater for making an "accurate record of diet" possible. However, Fisher was unsatisfied with the recommended practices of calorie counting. Atwater's tables were based on the weight of food and its percentage of fats, carbohydrates, and proteins. To determine how many calories were to be eaten, Fisher (1906, 417) complained, one had to

> weigh the food which is eaten, and to multiply the weight of each food by the three percentage figures . . ., and thus obtain the weight of proteid, fat, and carbohydrate; then to multiply the weight (in grams) of proteid and carbohydrate by the factor 4.1 in order to obtain the calorific value of these elements; and to multiply the weight of fat by the factor 9.3 in order to find the calorific value of fat. The results express the food value in "large calories" of proteid, fat, and carbohydrate, and the sum of these three is the total calorific value of the food.

This procedure was clearly too complicated and laborious for a life outside the laboratory—especially if one needed to determine the caloric content of various meals for a greater number of people, like Fisher's (1906, 417) friend John Harvey Kellogg did in his sanitarium in Battle Creek,

Michigan.[9] Not to count calories, however, was no option either: Fisher worried that people underestimated the amount of fuel and protein they consumed: Although people would often think small snacks such as nuts or popcorn did not matter, "[n]ature counts every calorie very carefully". If too many calories were consumed, the body would put on weight (Fisher and Fisk 1915, 32–3).

Fisher wrote these lines at a time when body fat was increasingly regarded as a questionable and antimodern sign of indulgence or illness (Schwartz 1986; Vester 2010; Mackert 2014b). Yet there was still another reason why not measuring one's food intake could lead to grave consequences: the alleged dangers of too much protein intake. Protein was a difficult case for many scientists who experimented on the efficiency of eating.[10] On the one hand, they regarded it as absolutely necessary to repair and build muscle. Whereas fats and carbohydrates seemed rather interchangeable in that they both provided fuel energy, for Fisher, the consumption of nitrogenous food was "absolutely essential" to avoid bodily deprivation and "ultimate . . . death" (Fisher and Fisk 1915, 36).[11] On the other hand, too much protein proved to be a problem. First, because protein was regarded as inefficient: Thermodynamic experiments had shown that bodily processes of burning proteins produced much more heat and, thus, much more potential heat loss or waste heat than the digestion of carbohydrates and fats (Neswald 2013, 91–2). Second, like many of his contemporaries, Fisher worried about distinct illnesses resulting from a diet too heavy in eggs and meat. He was convinced that this overstrained digestion and led to intestinal putrefaction and "autointoxication" (Fisher 1909, 42). The fear of autointoxication was widespread among nutritionists at that time. Many recommended not only a distinct diet but, most importantly, thorough mastication, so as to avoid this menace (Whorton 1982, 223–4; Levenstein 2012, sec. 3). Mastication was a method of slow chewing devised by Horace Fletcher, a mutual friend of both Fisher and Kellogg. By chewing every bite of food for several minutes, the proteins were meant to be absorbed already in the mouth, easing their digestion in the body, avoiding putrefaction, and bringing a great many other benefits (Fisher and Fisk 1915, 44–7; Whorton 1982, 205; Armstrong 1998, 43–4).[12]

For Fisher, autointoxication was an economic problem, too, because it led to fatigue and reduced work endurance. He believed "the economic waste from undue fatigue" to be "probably much greater than the waste from serious illness" (Fisher 1909, 47). In any case, fatigue increasingly concerned physiologists because it challenged the thermodynamic law of energy conservation. Obviously, there *were* ways of wasting energy, and fatigue was the factor that distinguished human motors from actual steam engines (Rabinbach 1990; Sarasin and Tanner 1999, 34; Vatin 1999). A controlled intake of protein, however, promised an increase of work endurance and efficiency, as Fisher (1907, 43) himself found in an experiment on the effects of thorough mastication combined with a low-protein diet (see Whorton

1982, 193). Thus, it was evident for him that people needed to carefully monitor their protein consumption (Fisher 1906, 429–30).

Particularly with regard to this important aspect, Fisher found Atwater's caloric tables highly "misleading". He criticized that measuring different foods such as milk and pecan nuts on the basis of "weight per cent" or "calories per ounce" led to the wrong impression that pecan nuts provided more "calories out of proteid" than milk, although the opposite was true. As a solution, Fisher suggested measuring foods by "calories per cent" and dividing them into standard portions of one hundred calories. Thereby, one could save the work of complicated calorie counting and immediately know the caloric concentration of protein in such a serving, making possible a "true comparison between different foods" (Fisher 1906, 417–19) such as "a small lamb chop (weighing about an ounce); . . . a large egg (about 2 ounces); . . . a very large dish of oatmeal (about 6 ounces); [and] a small piece of sponge-cake" (Fisher and Fisk 1915, 29). Whereas Fisher sought to improve calorie counting, Atwater used the calorie to develop another measurement of efficiency. As will be demonstrated in the following, his concept of a "pecuniary economy of food" provided a way to determine people's ability to consume productively.

Dietary Studies and the Problem of Productive Consumption

In 1887, when Atwater became head of the Office of Experiment Stations, an agency in the Department of Agriculture that coordinated the work of several newly established food research facilities in the country, he initiated a broad series of dietary studies all over the US (see Dirks and Duran 2001; Dirks 2003). Surveyors documented, for example, the diet of working-class people in New York City, of African Americans in the rural South, and of university boat crews (Atwater and Woods 1897; 1898; Atwater and Bryant 1900; 1902). They weighed all the food their test subjects were eating and measured its content in terms of protein, carbohydrates, and fats, as well as its caloric value. Atwater (1894) used the data to further develop caloric "dietary standards" for varying degrees of hard work. According to him, men "with little physical exercise" needed to consume 2,450 calories daily to maintain their physical status quo. For those "with hard muscular work", he even advised an intake of 5,700 calories.[13] Whereas Atwater's "dietary standards" were the object of controversy among nutritionists and were later revised by Atwater himself, one point was hardly doubted: the idea that calories could adequately express a causal relationship between food intake and energy expenditure, and, thus, work (see Chittenden 1905). This assumed correspondence between input and output seemed to permit an exact determination of the work that was possible to perform with a particular diet (Tanner 2010, 43). In his experiments with the calorimeter, for instance, Atwater sought to determine "how far a man ought to ride a bicycle on one egg".[14] Whereas this setting bespeaks the experimental character of nutritional knowledge, Atwater's studies and experiment did contribute to a highly virulent topic of reform: the problem of working-class nutrition.[15]

In his interpretation of the dietary studies, Atwater related men's ability to work, their ultimate productivity, to the amount of protein and calories they ate. A report on a painter's family in New York City, for instance, portrayed the family as always in debt with only "unsteady [e]mployment" and undernourished by a highly insufficient diet:

> The quantities of protein and energy consumed per man per day by this family were but little more than half the amount ordinarily considered necessary for men at moderate work, and undoubtedly more food than this would be required to maintain the family in good physical condition. A certain grade of work might of course be done upon a small amount of protein and energy; but it has been observed that within certain limits the character of the work improves as the diet becomes more liberal.
>
> (Atwater and Bryant 1902, 58)

The specific idea of productivity that the "human motor" promoted, namely the direct relationship of input and output, allowed Atwater to assume that the family members'—probably the men's—work and presumably also its outcome would increase by being better nourished. However, as Atwater stressed, not the family's limited budget per se, but unwise choices in spending this budget were responsible for the malaise. He recommended buying cheap carbohydrates such as oatmeal instead of green veggies and meat and promised, "Had the family spent their money for food [more] wisely, they would not have been so poorly nourished" (59). Hence, Atwater suggested that their capability to work and, thus, their employment, as well, would have perhaps been better and steadier had they been wiser regarding household budgeting.

Verdicts such as these came with important consequences for ongoing negotiations of workers' wages. Around 1900, food reformers had already for a while been highlighting the importance of eating "right" in the fight against poverty, economically and morally. Instead of raising workers' wages, as the newly established labor unions were urging, many Progressives wanted to teach the poor to "help themselves" to eat right and to efficiently manage their limited budgets (Atwater 1888, 444).[16] Atwater did not advocate higher wages either. For him, the problem of workers' nutrition was a problem of avoiding waste, of "getting the most out of what we have" (442).[17] The dietary studies seemed to confirm that there were many ways of improving workers' diets in regard to quantity and nutritional quality by demonstrating how they could better buy and cook their food.

For this purpose, Atwater developed a concept of food economy that was based on his caloric experiments. He defined food economy as buying and eating food "which is both most healthful and cheapest" (Atwater 1894, 10) and used the calorie to determine which foodstuffs allowed the most efficient "pecuniary economy of food" (1888, 442; see Neswald 2013, 101). The researchers who conducted the dietary studies had related the calories of foods to the price paid for them. This concept translated fuel value into monetary value with the intention of optimizing costs and consumption: How many calories could one buy for twenty-five cents? Which foodstuff came with the best ratio of fuel value and costs? In this way, Atwater came to such conclusions as substituting green vegetables for oatmeal because of the latter's bigger fuel value and lesser cost (see e.g., Atwater and Bryant 1902, 12, 59, 66). Health and low costs were declared to be the basic principles of food economy and the means and results of economically budgeting bodies. The solution to the problem of poverty was now to carefully select, buy, and consume specific foods.

In their focus on an improved consumption to achieve a higher productivity, the agendas of Fisher, Atwater, and other Progressive food reformers contributed to a significant development in American capitalism: Fueled by nutritional science, Progressive America saw a reconfiguration of consumption that changed its politics of inclusion and exclusion. In the decades around 1900 and especially in the early twentieth century, economic discourse shifted "from an analysis governed by scarcity, in which problems of production dominate[d], to one governed by abundance, in which consumption replaced production as the main focus" (Armstrong 1998, 53).[18] Already in the late nineteenth century, economists had critiqued a conservative emphasis on moral restraint and sought ways to legitimize increased consumption (Horowitz 1980, 301–02). In the early twentieth century, their efforts met with a Fordist rationality of accumulation that conceptualized production as geared toward consumption (Bänziger 2013, 14–15). Nutritional science emerged from and built upon this reconfiguration of consumption. As an effect of the thermodynamic model of the body, nutritional knowledge legitimized and even necessitated consumption.[19] Atwater's dietary studies stressed that workers should not be limited in their daily needs because, otherwise, their ability to work would suffer. Conceptualizing food as fuel for the human motor and its input as determining the output of work energy, the Puritan adage of "he who does not work shall not eat" could be reversed into "he who has nothing to eat cannot work" (Tanner 2010, 42). Atwater even acknowledged that food had to satisfy one's palate and thus recommended consuming the largest variety of foodstuffs possible (Atwater 1888, 442; see e.g., Atwater and Bryant 1902, 17, 30, 67).

Although nutritional reform encouraged consumption and thus included the poor, in a certain way, in consumer society, this movement surely did not allow for food intake to go unchecked. Not only should diets be adjusted to people's budgets, but reformers also issued strong warnings against overeating. Thus, the idea of healthy, economical food mediated the tension between consumption and productivity by identifying beneficial wants and, nevertheless, calling for the necessary restraint (see e.g., Horowitz 1980, 302). The calorie and the concept of food economy worked as measurements of such a productive consumption, as budgeting emerged as a condition and marker of people's ability to consume "right"—that is, to be geared toward productivity. This ability was an integral part of the regime of able-bodiedness, because it was closely tied to the corporeal. By relating poor people's ability to work to their budgeting, the nexus of nutritional science and Progressive reform ethos created a new ideal of bodies that were buying and consuming properly to enable and maintain their ability to work.

Thus, the turn to consumerism changed but also maintained the politics of exclusion—a dynamic that can also be described with regard to the racial politics of food reform. Interestingly enough, the consumerist orientation of nutritional science challenged late nineteenth century understandings of race and racial progress (Veit 2013, sec. 5). The fear of "miscegenation" and the increasing power of the "one-drop rule" illustrate the strong emphasis that eugenicists placed on sexual reproduction and genetic succession (Davis 1991; Hodes 1997). Fisher himself highlighted the important role of heredity and sexual hygiene in the book *How to Live*, which he had written together with physician Eugene L. Fisk in 1915, and that book became a huge success.[20] Fisher and Fisk (1915, xv–xix) were both members of the Life Extension Institute, a coalition of economists, eugenicists, and food reformers who sought to promote health and longevity. However, whereas the authors reminded their readers that they were "the trustees of the racial germ plasm that we carry" and "under the most solemn obligation to keep it up to the highest level", they suggested ways to do that which by far exceeded heredity and proved a more fluid understanding of racial progress—such as healthy foods, clean air, and physical activity (165). If the consumption of healthy food allowed for real improvement, the "survival of the fittest" wasn't solely a question of genes.

This consumerist extension of eugenics was expressed in the label "euthenics" that chemist and home economist Ellen Richards coined around 1910. Richards, too, sought to improve the efficiency of eating and explained the benefits of euthenics in contrast to eugenics: "Eugenics deals with race improvement through heredity. Euthenics deals with race improvement through environment. Eugenics is hygiene for the future generations. Euthenics is hygiene for the present generation. Eugenics must await careful investigation. Euthenics has immediate opportunity" (Richards [1910] 1912, viii).[21] By quoting Fisher's estimates of the economic loss by preventable deaths, Richards further justified this immediate effort and inscribed her agenda into the comprehensive program of individual and national improvement. It is important to note that although euthenic thought displayed a changed understanding of race and highlighted its fluidity, it wasn't less productive in its rationale for racial uplift and its inscription of racial differences.

Moreover, euthenics came with similar dynamics of inclusion and exclusion, because it promised not only *survival* but also *improvement* of the fittest. "There is a still higher stage of medical science than that of fighting or preventing disease in the individual", Fisher (1909, 15) promised the readers of the report, "the stage which has been called 'biological engineering,' i.e., the study of the conditions under which the individual may reach his highest efficiency". Describing this endeavor as a promising new road to true human freedom that abandons outdated traditions, Fisher marked the constant process of improvement as ultimately modern (see e.g., Fisher and Fisk 1915, 1–2). He condemned his fellow economists for "fixing attention exclusively on physical phenomena," thereby "leav[ing] out of [the] account the most essential element of all, the vigor of human life". For him, "the true 'wealth of nations' [was] the health of its individuals" (Fisher [1906] 1919, 176).[22] It was this positive, open-ended ideal of the improvability of bodies by thorough consumption that constituted the distinctly consumerist character of able-bodiedness in the Progressive Era. To trace its effects and to examine its markedly producerist edge, it is helpful to have a closer look at Fisher's concept of human capital.

The Income Value of Able-Bodiedness

Fisher's work on capital theory is known for understanding and quantifying humans as national assets (Dimand 2005; Nordhaus 2005).[23] For him, productivity was the central feature of capital: "Capital is that wealth which yields income" (Fisher [1906] 1919, 184). Capital differed from wealth by its capacity to produce output, in this case, income. Human capital then, which Fisher explicitly included in his capital definitions, was to be measured by income—and that was the figure that Fisher (1897, 535) used for the calculation of the annual losses from illnesses and death for *National Vitality*.[24] Interestingly enough, Fisher further differentiated this "capitalization of earning power" into "objective" and "subjective income". A person's objective income consisted of his or her actual budget for food, housing, etc. The subjective income, however, included the subjectively felt ordeal of work. It is most interesting for an analysis of contemporary understandings of productive bodies, because Fisher ([1906] 1919, 168) closely tied earnings to bodies and their well-being. "Objective income takes no account of the toil of the laborer", he explained:

> The workingman who earns $2 a day earns double the objective income of one who earns $1 a day. Yet, if the work of the former is difficult, loathsome, or dangerous, it may well be that many would prefer the nominally smaller income rather than endure these disadvantages.
> (170)

Thus, a worker's objective income could be lessened not only by the hardship of labor but also by the condition of the worker, "by an abnormal state of the body" (Fisher [1906] 1919, 178), by pains and illnesses. Whereas, for Fisher (171), the toil of work was a necessary evil rather than subject to reform, he deemed individual health and capabilities to be the foundation of improvement: "A man with a good constitution has . . . more . . . subjective income . . . than one without" (176). Because he saw a causal link between a distinct diet and health, he regarded healthy eating as an "investment into the body" (170) that increased the subjective income with interest rates of oftentimes "several thousand per cent" (Fisher 1909, 12). For Fisher, bodies were key to the maximization of income. Consequently, a few years later, he was no longer simply talking about humans as "vital assets" but emphasizing their future potential. In *How to Live*, he envisioned a "vital surplus [that] should not only be safeguarded, but accumulated" as "balance in the savings bank of life" (Fisher and Fisk 1915, sec. 5).[25]

His vision probably sounded promising not only to the ears of contemporary labor physiologists who feared fatigue and the limits of human capabilities. By capitalizing on bodily abilities and consumption, Fisher's capital theory built on and strengthened a compulsory able-bodiedness that was a powerful ideal for a consumerist society and fueled the turn to consumerism discussed earlier. However, as Fisher's interventions indicate, this emphasis on consumption did not supersede a focus on production.[26] It continued to have a producerist character, because it was conceptualized as a vehicle for increased productivity and the enhancement of life. Fisher and Fisk (1915, 1) described the benefits of "individual hygiene" in a comprehensive racial, social, and moral agenda as leading to "health, strength, endurance, symmetry, and beauty; [increasing] our capacity to work, to be happy, and to be useful; [developing] not only the body, but [also] the mind and the heart; [ennobling] the man as a whole".

Central to this ideal of improvement was that the latter was self-made. Fisher and Fisk (1915, 3) welcomed and encouraged efforts "by the people themselves" to eat healthily and to promote regulation for more hygienic and healthy food and drinks. People's self-conduct became the focus of reform efforts—for the experts as well as for the people themselves—and this came with the important imperative to avoid misconduct in any way. Fisher and Fisk admitted that the old "ideals of health" had allowed for minor transgressions, because they had simply aimed at avoiding sickness. The new understanding of "good health in the sense of attaining the highest physical and mental efficiency and power" was "utterly inconsistent" with any of those breaches (viii). A constitutive part of the comprehensive norm of able-bodiedness was the ability of successful self-conduct.

Prospect: Calorifying Eating Practices

In light of the history of nutritional science and calories in the Progressive-Era US, it is not difficult to see how the *dispositif* of diet and nutrition constituted a point of intersection between regulating the individual and populations, as Foucault (1979; 1988) has argued for sexuality (see Möhring 2012). In the foreword to *How to Live*, former US president William Howard Taft described health in a profoundly biopolitical manner as "the foundation on which reposes the happiness of the people and the power of a country", hailing "the care of individual and family health [as] the first and most patriotic duty of a citizen" (Fisher and Fisk 1915, iv).[27] One of the distinct techniques of this governmentality was the calorie, as a brief return to Fisher's one-hundred-calorie portions illustrates.

To simplify calorie counting and to provide a quick overview on the nutritional content of food, the economist developed a device that visualized the nutritional composition of meals even when different foods were consumed together (Fisher 1906). It would be exaggerative to say that Fisher's method was simple. It involved a bolted metal apparatus, cardboards that were to be clamped into that, and heavy-headed pins to be pierced into the card, indicating the number of servings and producing a center of gravity that displayed the nutritional ratio (427–9). However, for Fisher, it was superior, especially in its visualization of caloric content and nutrients at the same time. Once an individual was made familiar with the new method, Fisher hoped that the apparatus would not be needed further, as the graphic display of food composition would have made its way into the subject's visual memory (431).

Whereas Fisher's apparatus apparently did not have a great future, the calorie did, by occupying the intersection between shaping food politics and the dietary practices of individuals. In 1918, the physician Lulu Hunt Peters (1918, 24), credited with writing the first diet manual based on calorie counting, urged her readers to change the way they were understanding their food: "Instead of saying one slice of bread, or a piece of pie, you will say 100 Calories of bread, 350 Calories of pie", she demanded, and made clear, "Hereafter you are going to eat calories of food". Obviously, the calorie should become a mode for dieters to understand their food—and themselves. It increasingly provided a way in which they could work on themselves and demonstrate their ability to consume right. The one-hundred-calorie portions (that were taken up by Peters and others in the following decades and are still shaping dietary practices today) illustrate that the career of the calorie should continue in the course of the twentieth century.

The symbolic power of "investments into the body" and the relationship between one's health and productivity resonated in a variety of contexts. The nexus of nutritional science and economics was constitutive in not only shaping concepts of productivity and consumerism but also in mediating the tension between those rationales—by contributing to the establishment of able-bodiedness as a norm of consumerist productivity. The moral imperative of this *dispositif* was strong, because allegedly neglecting the rules and practices of healthy eating meant disregarding individual and public health as well as life itself. The calorie distributed and inscribed the knowledge and ideal of eating "right" into individual routines. Calorie counting echoed the idea of a personal responsibility for one's health, and because of the future-oriented promise of healthy eating, this measuring became an obligation that was basically endless. This is a central marker of compulsory able-bodiedness, as it incites the need of continuously caring for one's body and self. In this vision of fruitful and ever-more efficient futurity, compulsory able-bodiedness contributes to the "regime of productivity" of the twentieth century. To this day, calories are one of the means by which this regime is acted out. The history and practice of calorie counting thus reveal the political character of food and eating.

Notes

1 I am grateful to Peter-Paul Bänziger and Mischa Suter as well as to Felix Krämer, Jürgen Martschukat, and Cécile Stehrenberger for their instructive comments on earlier versions of this chapter.

2 My understanding of productivity as an analytical term builds on the work of Michel Foucault (1980; 1988) and Ernesto Laclau (1990, 2005) and highlights the capacity of privileged signifiers to produce knowledge, material orders, and subjects (Mackert 2012; 2014a).

3 Regarding Fisher, see Robert W. Dimand and John Geanakoplos (2005), James C. Whorton (1982, 191–7) and Robert Loring Allen (1993).

4 Robert McRuer (2002, 88–99; 2006) refers to Adrienne Rich's concept of "compulsory heterosexuality", not only identifying parallels between the norms but also considering them as densely interwoven.

5 Regarding the concept of *dispositif*, see Michel Foucault (1980).

6 Robert McRuer (2002, 90) argues, "it is precisely the introduction of normalcy into the system that introduces compulsion".

7 Regarding the perspective of Critical Ability Studies, see Nina Mackert (2014b).

8 For a more detailed account of Atwater's work with the calorimeter, see Jessica Mudry (2009, 31–9) and Nick Cullather (2007).

9 Regarding Kellogg and the sanitarium, see Nina Mackert (forthcoming), Gerald Carson (1957), and Adam D. Schprintzen (2013, sec. 5).

10 Regarding the history of protein, see Kenneth John Carpenter (1994).

11 Some contemporaries thus argued that the increased consumption of meat was a sign of civilizational progress. See Nina Mackert (2014b, 228–9).

12 Regarding the racial and gendered politics of autointoxication and mastication, see R. Marie Griffith (2000). Fisher, Kellogg, and Fletcher were all members of the Health and Efficiency League of America, a club whose name bespeaks their common interest (Whorton 1982, 195; Schwartz 1986,186).

13 These standards more than hint at the fact that, not surprisingly, the prototype of the human motor was male. Although Atwater's staff also measured the food intake of wives, other female relatives, and children, they counted their meals only as fractions of male family members' meals. Regardless of what they had eaten and regardless as to whether they did wage work, housework, or the like, the meals of females were counted as 0.8 times the meals of men (see Atwater and Bryant 1902). Such dietary accounts reified and naturalized gender distinctions and a hierarchical access to food for men and women (see Tanner 1999, 77). For the highly gendered character of nutritional science at the *fin de siècle*, see also Maria Osietzki (1999).

14 "Cyclist to Aid Cause of Science". *Chicago Daily Tribune*, A4. October 30, 1904; quoted in Nick Cullather (2007, 340).

15 I am indebted to Felix Krämer for his inspiring cooperation on this topic; this led to a joint talk on "Budgeting Bodies: Home Economics and the Space of Class"

at the Conference "Histories of American Foodways" of the German Association for American Studies, in Erfurt, February 2015.

16 Regarding the efforts of food reformers, see Laura Shapiro (1986), Harvey A. Levenstein ([1988] 2003, sec. 4, 6), Charlotte Biltekoff (2013, sec. 2).

17 Regarding the role of the fear of waste in twentieth century productivity discourse, see Lukas Held, this volume.

18 See also Jackson Lears (1994, 10–11), Lizabeth Cohen (2003, 21–2) as well as Heiko Stoff (2001, 446); the latter is referring to Lawrence Birken (1988).

19 Maria Osietzki (1999, 343) suggests that thermodynamic thought influenced the turn to consumerism, because it legitimized consumption and was opposed to excessive scarcity and restraint in one's diet.

20 The book was published in nineteen editions and sold approximately four hundred thousand copies by 1937 (Dimand 2005, 393).

21 Regarding Richards, see Harvey A. Levenstein ([1988] 2003, sec. 4, 6), Charlotte Biltekoff (2013, sec. 2) and Laura Shapiro (1986).

22 Of course, Fisher was referring to Adam Smith's (1776) seminal work *An Inquiry into the Nature and Causes of the Wealth of Nations.*

23 Similar to human capital theory were ideas of human resource management that had been discussed and applied at workplaces since 1870 (see Kaufman 2008).

24 Building on earlier works by the British economist William Farr as well as census data, Fisher (1909, 118–19) estimated that "the average economic value of the inhabitants of the United States . . . is $2,900", thereby calculating the value of the "vital assets" of the nation to be $250 billion.

25 Fisher's promise of the enhancement of human capital hints at a shift that Lukas Held (this volume) stated for the second half of the twentieth century when the focus on conservation of bodily resources gave way to a future-oriented vision of ever-increasing efficiency.

26 For a similar argument of consumerism as supplementing, not substituting, producerism, see Peter-Paul Bänziger (2015; 2013, 16).

27 The first quote is a quote from former British prime minister Benjamin Disraeli, which Taft used here. It is remarkable that Taft was in a position to say that because his girth had prompted the New England Fat Men's Club to invite him to become a member (Mackert 2014b, 230). However, Taft had spent a few years dieting again and again (Sotos 2003) and in the eyes of his contemporaries, this might have been sufficient to prove his ability of adequate self-conduct. For the discursive link between dieting and self-control, see Katharina Vester (2010).

Bibliography

Allen, Robert Loring. 1993. *Irving Fisher: A Biography*. Hoboken: Wiley-Blackwell.

Armstrong, Tim. 1998. *Modernism, Technology, and the Body: A Cultural Study*. Cambridge, UK: Cambridge University Press.

Atwater, Wilbur O. 1888. "The Pecuniary Economy of Food." *The Century Magazine* 35: 437–46.

———. 1894. "Foods: Nutritive Value and Costs." *Farmer's Bulletin of the U.S. Department of Agriculture* 23.

——— and A.P. Bryant. 1900. "Dietary Studies of University Boat Crews." *Bulletin of the U.S. Department of Agriculture, Office of Experiment Stations* 75.

——— and A.P. Bryant. 1902. "Dietary Studies in New York City in 1896 and 1897." *Bulletin of the U.S. Department of Agriculture, Office of Experiment Stations* 116.

——— and Chas D. Woods. 1897. "Dietary Studies with Reference to the Food of the Negro in Alabama." *Bulletin of the U.S. Department of Agriculture, Office of Experiment Stations* 38.

——— and Chas D. Woods. 1898. "Dietary Studies in New York City in 1895 and 1896." *Bulletin of the U.S. Department of Agriculture* 46.
Bänziger, Peter-Paul. 2013. "Fordistische Körper in der Geschichte des 20. Jahrhunderts—eine Skizze." *Body Politics* 1 (1): 11–40.
———. 2015. "Von der Arbeits- zur Konsumgesellschaft. Kritik eines Leitmotivs der deutschsprachigen Zeitgeschichtsschreibung." *Zeithistorische Forschungen/ Studies in Contemporary History* 12: 11–38.
Bederman, Gail. 1995. *Manliness & Civilization: A Cultural History of Gender and Race in the United States, 1880–1917*. Chicago: University of Chicago Press.
Biltekoff, Charlotte. 2013. *Eating Right in America: The Cultural Politics of Food and Health*. Durham: Duke University Press.
Birken, Lawrence. 1988. *Consuming Desire: Sexual Science and the Emergence of a Culture of Abundance, 1871–1914*. Ithaca: Cornell University Press.
Butler, Judith. 1991. "Imitation and Gender Insubordination." In *Inside/Out: Lesbian Theories, Gay Theories*, edited by Diana Fuss, 13–31. New York: Routledge.
Carpenter, Kenneth John. 1994. *Protein and Energy: A Study of Changing Ideas in Nutrition*. Cambridge: Cambridge University Press.
Carson, Gerald. 1957. *Cornflake Crusade*. New York and Toronto: Rinehart.
Chittenden, Russell H. 1905. "Economy in Food." *The Century Magazine* 70: 859–70.
Cohen, Lizabeth. 2003. *A Consumers' Republic: The Politics of Mass Consumption in Postwar America*. New York: Alfred A. Knopf.
Cullather, Nick. 2007. "The Foreign Policy of the Calorie." *The American Historical Review*, 112 (2): 337–64.
Dairy Council of California. n.d. "Correct Portion Sizes: How to Keep Portion Distortion in Check." Accessed November 30, 2015. http://www.healthyeating.org/Healthy-Eating/Healthy-Living/Weight-Management/Article-Viewer/Article/348/correct-portion-sizes-how-to-keep-portion-distortion-in-check.aspx.
Davis, F. James. 1991. *Who Is Black? One Nation's Definition*. University Park: Penn State University Press.
Dimand, Robert W. 2005. "Comments on William D. Nordhaus's, 'Irving Fisher and the Contribution of Improved Longevity to Living Standards'." *American Journal of Economics and Sociology* 64 (1): 393–7.
——— and John Geanakoplos, eds. 2005. "Celebrating Irving Fisher: The Legacy of a Great Economist." Special issue of *The American Journal of Economics and Sociology* 64 (1).
Dirks, Robert T. 2003. "Diet and Nutrition in Poor and Minority Communities in the United States 100 Years Ago." *Annual Review of Nutrition* 23: 81–100.
——— and Nancy Duran. 2001. "African American Dietary Patterns at the Beginning of the 20th Century." *Journal of Nutrition* 131: 1881–9.
Fisher, Irving. 1897. "The Role of Capital in Economic Theory." *The Economic Journal* 7 (28): 511–37.
———. 1906. "A New Method for Indicating Food Values." *American Journal of Physiology* 15: 417–32.
———. 1907. *The Effect of Diet on Endurance*. New Haven: Tuttle, Morehouse & Taylor Press.
———. 1909. *National Vitality, Its Wastes and Conservation, Report of the National Conservation Commission*. Washington, DC: Government Printing Office.
———. (1906) 1919. *The Nature of Capital and Income*. New York: The Macmillan Company.

——— and Eugene Lyman Fisk. 1915. *How to Live: Rules for Healthful Living based on Modern Science*. New York and London: Funk and Wagnalls.
Foucault, Michel. 1979. *Discipline and Punish: The Birth of the Prison*. Translated by Alan Sheridan. New York: Random House.
———. 1980. "The Confession of the Flesh (1977), Interview." In *Power/Knowledge: Selected Interviews and Other Writings, 1972–1977*, edited by Colin Gordon, 194–228. New York: Pantheon.
———. 1988. *The History of Sexuality, Vol. III: The Care of the Self*. Translated by Robert Hurley. New York: Pantheon Books.
Griffith, R. Marie. 2000. "Apostles of Abstinence: Fasting and Masculinity During the Progressive Era." *American Quarterly* 52 (4): 599–638.
Hargrove, James L. 2006. "History of the Calorie in Nutrition." *The Journal of Nutrition* 136 (12): 2957–2961.
Hodes, Martha. 1997. *White Women, Black Men: Illicit Sex in the Nineteenth-Century South*. New Haven: Yale University Press.
Horowitz, Daniel. 1980. "Consumption and Its Discontents: Simon N. Patten, Thorstein Veblen, and George Gunton." *Journal of American History* 67 (2): 301–17.
Kaufman, Bruce. 2008. *Managing the Human Factor: The Early Years of Human Resource Management in American Industry*. Ithaca: Cornell University Press.
Laclau, Ernesto. 1990. *New Reflections on the Revolutions of Our Times*. London: Verso.
———. 2005. *On Populist Reason*. London: Verso.
Lears, Jackson. 1994. *Fables of Abundance: A Cultural History of Advertising in America*. New York: Basic Books.
———. 2009. *Rebirth of a Nation: The Making of Modern America, 1877–1920*. New York: Harper Collins.
Levenstein, Harvey A. (1988) 2003. *Revolution at the Table: The Transformation of the American Diet*. Berkeley: University of California Press.
———. 2012. *Fear of Food: A History of Why We Worry about What We Eat*. Chicago: University of Chicago Press.
Mackert, Nina. 2012. "'But Recall the Kind of Parents We Have to Deal With': Juvenile Delinquency, Interdependent Masculinity and the Government of Families in the Postwar U.S." In *Inventing the Modern American Family. Family Values and Social Change in 20th Century United States*, edited by Isabel Heinemann, 196–219. Frankfurt am Main: Campus.
———. 2014a. *Jugenddelinquenz: Die Produktivität eines Problems in den USA der späten 1940er bis 1960er Jahre*. Konstanz: UVK.
———. 2014b. "'I Want to Be a Fat Man / And with the Fat Men Stand'—U.S.-Amerikanische Fat Men's Clubs und die Bedeutungen von Körperfett in den Dekaden um 1900." *Body Politics* 2 (1): 215–43.
———. Forthcoming. "Making Food Matter: 'Scientific Eating' and the Struggle for Healthy Selves." In *Food, Power, and Agency*, edited by Jürgen Martschukat and Bryant Simon. New York: Bloomsbury.
McRuer, Robert. 2002. "Compulsory Able-Bodiedness and Queer/Disabled Existence." In *Disability Studies: Enabling the Humanities*, edited by Rosemarie Garland-Thomson, Brenda Jo Brueggemann, and Sharon L. Snyder, 88–99. New York: MLA Publications.
———. 2006. *Crip Theory: Cultural Signs of Queerness and Disability*. New York: New York University Press.

Megan, Patrick. n.d. "What Does 100 Calories Look Like? Learn How to Estimate Portion Sizes." *SparkPeople*. Accessed November 30, 2015. http://www.sparkpeople.com/resource/nutrition_articles.asp?id=1987.
Möhring, Maren. 2012. "Essen." In *What Can a Body Do? Praktiken und Figurationen des Körpers in den Kulturwissenschaften*, edited by Netzwerk Körper, 47–56. Frankfurt: Campus.
Mudry, Jessica. 2009. *Measured Meals: Nutrition in America*. Albany: SUNY Press.
Neswald, Elizabeth R. 2013. "Kapitalistische Kalorien. Energie und Ernährungsökonomien um die Jahrhundertwende." In *Szenarien der Energie. Zur Ästhetik und Wissenschaft des Immateriellen*, edited by Barbara Gronau, 87–108. Bielefeld: transcript.
Nordhaus, William D. 2005. "Irving Fisher and the Contribution of Improved Longevity to Living Standards." *American Journal of Economics and Sociology* 64 (1): 367–92.
Osietzki, Maria. 1999. "Körpermaschinen und Dampfmaschinen. Vom Wandel der Physiologie und des Körpers unter dem Einfluß von Industrialisierung und Thermodynamik." In *Physiologie und Industrielle Gesellschaft. Studien zur Verwissenschaftlichung des Körpers im 19. und 20. Jahrhundert*, edited by Philipp Sarasin and Jakob Tanner, 313–46. Frankfurt am Main: Suhrkamp.
Peters, Lulu Hunt. 1918. *Diet and Health with Key to the Calories*. Chicago: The Reilly & Lee Co.
Rabinbach, Anson. 1990. *The Human Motor: Energy, Fatigue, and the Origins of Modernity*. New York: Basic Books.
Richards, Ellen H. (1910) 1912. *Euthenics: The Science of Controllable Environment*. Boston: Whitcomb & Barrows.
Sarasin, Philipp, and Jakob Tanner. 1999. "Physiologie und industrielle Gesellschaft: Bemerkungen zum Konzept und zu den Beiträgen dieses Sammelbandes." In *Physiologie und Industrielle Gesellschaft. Studien zur Verwissenschaftlichung des Körpers im 19. und 20. Jahrhundert*, edited by Philipp Sarasin and Jakob Tanner, 12–43. Frankfurt am Main: Suhrkamp.
Schprintzen, Adam D. 2013. *The Vegetarian Crusade: The Rise of an American Reform Movement, 1817–1921*. Chapel Hill: University of North Carolina Press.
Schwartz, Hillel. 1986. *Never Satisfied: A Cultural History of Diets, Fantasies, and Fat*. New York: Anchor Books.
Shapiro, Laura. 1986. *Perfection Salad: Women and Cooking at the Turn of the Century*. New York: Farrar Straus and Giroux.
Simon, Bryant. 2003. "'New Men in Body and Soul': The Civilian Conservation Corps and the Transformation of Male Bodies and the Body Politic." In *Seeing Nature Through Gender*, edited by Virginia J. Scharff, 80–102. Lawrence: University Press of Kansas.
Smith, Adam. 1776. *An Inquiry into the Nature and Causes of the Wealth of Nations*. London: Methuen & Co.
Sotos, John G. 2003. "Taft and Pickwick. Sleep Apnea in the White House." *Chest Journal* 124 (3): 1133–142.
SparkPeople. n.d. "About Us." *SparkPeople*. Accessed November 30, 2015. http://www.sparkpeople.com/about/index.asp.
Stoff, Heiko. 2001. "Utopian Thinking between Producerism and Consumerism: What Distinguished the American New Deal from the German Volksgemeinschaft?"

In *Visions of the Future in Germany and America*, edited by Norbert Finzsch and Hermann Wellenreuther, 445–67. Oxford: Berg.

Tanner, Jakob. 1999. *Fabrikmahlzeit. Ernährungswissenschaft, Industriearbeit und Volksernährung in der Schweiz 1890–1950*. Zurich: Chronos.

———. 2010. "Lebensmittel und neuzeitliche Technologien des Selbst: Die Inkorporation von Nahrung als Gesundheitsprävention." In *Das präventive Selbst: Eine Kulturgeschichte moderner Gesundheitspolitik*, edited by Martin Lengwiler and Jeanette Madarasz, 31–54. Bielefeld: Transcript.

Tichi, Cecilia. 1987. *Shifting Gears: Technology, Literature, Culture in Modernist America*. Chapel Hill: University of North Carolina Press.

Treitel, Corinna. 2008. "Max Rubner and the Biopolitics of Rational Nutrition." *Central European History* 41 (1): 1–25.

USDA. n.d. "Eat the Right Amount of Calories For You." *United States Department of Agriculture*. Accessed November 30, 2015. http://fnsweb01.edc.usda.gov/weight-management-calories/weight-management/better-choices/amount-calories.html.

Vatin, Francois. 1999. "Arbeit und Ermüdung. Entstehung und Scheitern der Psychophysiologie der Arbeit." In *Physiologie und Industrielle Gesellschaft. Studien zur Verwissenschaftlichung des Körpers im 19. und 20. Jahrhundert*, edited by Philipp Sarasin and Jakob Tanner, 347–68. Frankfurt am Main: Suhrkamp.

Veit, Helen Zoe. 2013. *Modern Food, Moral Food: Self-Control, Science, and the Rise of Modern American Eating in the Early Twentieth Century*. Chapel Hill: University of North Carolina Press.

Vester, Katharina. 2010. "Regime Change: Gender, Class, and the Invention of Dieting in Post-Bellum America." *Journal of Social History* 44 (1): 39–70.

Whorton, James C. 1982. *Crusaders for Fitness: The History of American Health Reformers*. Princeton: Princeton University Press.

Wiebe, Robert H. 1967. *The Search for Order, 1877–1920*. New York: Hill and Wang.

8 Regaining Sufficiency

Work Therapy in 1930s German Internal Medicine[1]

Alexa Geisthövel

In 1932, two physicians from the Medical Clinic of Greifswald University published a book entitled *Produktive Diabetikerfürsorge* (Productive Diabetic Relief). They reported on the activities of a treatment facility for diabetics that had been established in 1930 in rural Garz, on the island of Rügen (Lublin and Kroner 1932). The report advocated a training program for patients suffering from different types of pancreas-related metabolic diseases. In the past, they had been written off as "metabolic cripples" (Katsch 1937, 401); now they would become people capable of earning a living, going to school, or keeping house. A key element of this transformation was physical exercise or, more precisely, Arbeitstherapie (work therapy). In so doing, the protagonists of the Garz Diabetikerheim (diabetics' home) explicitly linked a rather unspecific notion of productivity to the working body.

Garz was not the only place where "social medicine" stepped in to offer remedies for individual illnesses that contemporaries discussed as having serious ramifications on the welfare-state economy. Besides Garz, this paper examines the Work Therapy Department that was established in the 1930s at the Heidelberg Medical Clinic. Viktor von Weizsäcker, head of the department, never labeled the treatment it offered as being "productive". Still, its activities revolved around problems regarding work capability and performance as related to the legal and practical framework of contemporary social insurance or welfare.

Work therapy's scheme for producing productive bodies participated in a broad, historical process that can be described as the "(re-)production of work and work society" (Bänziger 2015), which goes back to the late nineteenth century. The two examples show how, within this general context, work-related internal medicine abandoned one of the formative approaches of interwar internal medicine, the "rationalization" of the human motor, for a different model: the maintenance of adaptable functions within a living "system". This reorientation is captured in a small, but significant semantic shift that substituted the overall formula for optimizing, "efficiency" with the less demanding "sufficiency".

Work Therapy in Interwar Germany: Two Examples

The history of work therapy is closely related to psychiatric institutions, where, since the early nineteenth century, "work" or "occupation" had been considered to be therapeutically useful. Generally, this type of treatment focused on maintaining discipline in an environment that was characterized by monotony and a potential for conflict. Often the therapeutic outlook barely covered its exploitative character (see Ankele and Brinkschulte 2015). At the beginning of the twentieth century, however, a more extroverted type of psychiatric work therapy emerged, which was meant to prepare patients for professional life outside the medical institution. A model that rose to prominence in the mid-1920s was psychiatrist Hermann Simon's *aktivere Krankenbehandlung* (more active treatment; see Germann 2007).

Other medical subdisciplines transformed their tradition of "exercise" into work therapy as well. In the 1930s, surgeon Karl Gebhardt (1937), pioneering work therapist and SS-physician sentenced to death at the Nuremberg trials, offered courses teaching his method to interns from different medical disciplines.[2] In Berlin, hospital managers were mapping out a centralized "exercise department" for 1,500 convalescents from different clinics, whereas clinicians argued that work therapy should stay in the social framework of the ward, directed by the attending physician (Trachte 1935; Dennig and Becker-Freyseng 1938). Work therapy was compatible with the radical eugenicist rationale of "extinction and healing" (Ebbinghaus and Dörner 2001) as well as with leftist "social therapy" (Unger 1931; Weizsäcker 1949, 92). However, as much as these approaches differed, they all shared the conviction that there was a blank space in the procedures of medical treatment that must be filled with "therapy towards work through work" (Siebeck 1934, 888). This is also true for the two examples to be examined here.

The Garz diabetics' home started as a public charity of the Inner Mission. In 1928, when its founder, Gerhardt Katsch, became director of the Medical Clinic of Greifswald University, the hospital offered no resources for a separate, off-site clinic that allowed patients to become accustomed to the complex diabetes treatment under conditions resembling "real life"—which was the rationale of "productive relief". After 1936, the board of the home was taken over by the university in order to establish a research institute. In 1939, the regional Landesversicherungsanstalt (State Insurance Institution) offered to sponsor a clinic extension worth 1.25 million reichsmark in exchange for the presidency of the trust, which Katsch welcomed.[3] His diabetics' home proved to fit three different political systems: the Weimar republic, the Nazi period, and after 1945, the enlarged institution would become one of the cornerstones of diabetes treatment in the German Democratic Republic.

The Garz home would not have come into existence without the market launch of insulin that transformed today's type 1 diabetes from a fatal to a chronic disease (see Feudtner 2003). This new therapeutic option required different coping strategies and put much emphasis on patient training and education. Next to diet and insulin injections, muscular activity worked as a direct remedy, because it helped the metabolic system to utilize carbohydrates, which consequently brought down the amount of insulin needed to counterbalance sugar levels. In Katsch's scheme, muscular activity, or work therapy, thus allowed for health cost savings, as well as for the metabolic and emotional stability of patients suffering from fairly serious diabetes.

The Garz home was supposed to represent the economic effects of diabetes treatment en miniature. The staff members—nurses, cooks, maids, lab assistants, and one of the physicians—were diabetics themselves and many of them former patients. They embodied the possibility of doing "normal" work, with the home creating paid positions for them (Katsch 1930; 1942; Banse and Spickernagel 1939, 59–65). Also, through work therapy, Garz became a self-supporting household. The residents grew and bred a substantial part of their own food and, thus, reduced the costs of treatment.

In this respect, Garz differed from the second example, the Medical Clinic of Heidelberg University. Work therapy evolved at the neurological department, headed by Viktor von Weizsäcker, who is today renown as a theorist of medical anthropology. In February 1930, he set up a special ward for the treatment of neurotic "working insufficiency". The Neuroses Department focused on the ailment of "pension neurosis", which physicians and social politicians at that time discussed as resulting from social security provisions, especially occupational accident insurance and military service compensation. According to Weizsäcker, pension neurotics were not trying to evade work at employers' or the public's expense, but were seeking legal and social justice. Thus, he preferred the term "law neurosis". He thought the patient must be convinced to drop all pension claims and go back to work, but only after adequate treatment, which included settling lawsuits and personal conflicts in the family or at the workplace (Weizsäcker 1930, 80–1).

In 1934, the Neuroses Department was closed due to lack of personnel and relaunched in late 1937, now under the official name of the Work Therapy Department. This resulted from a coordinated effort on the part of the directors of the Surgical, the Orthopedic, and the Medical Clinics.[4] The new department was designed for a specific phase of illness when medical findings were negative, but the patient was still recovering. It addressed convalescents of different internal affections and was situated outside the city on the grounds of the Orthopedic Clinic, where the patients could take walks and use rehabilitation facilities. The patients lived together with some degree of self-organization, including meals and cleaning. Just like Katsch's diabetics' home, this external department served as a bridge between clinic and life outside. Yet, in contrast with Garz, Weizsäcker (1939, 50–1) and his associates held that work done for therapeutic reasons must not serve any economic or technical purpose such as training someone for a specific job.[5] In August 1939, the department had to shut down when the entire Orthopedic Clinic was turned into a military hospital.[6]

Whereas the development of work therapy in the Heidelberg special wards suffered several setbacks, the Garz diabetics' home operated continuously. It achieved a strong reputation and, in the beginning of 1939, could look back on well over three thousand cases of treatment. In contrast, the Neuroses and Work Therapy Departments treated about seven hundred patients altogether.[7] As will be demonstrated, the two examples had many features in common, not least of which was their conservative but at the same time enabling approach. Katsch and Weizsäcker represented two different versions of "mainstream holism" in medicine, which strove to combine the achievements of natural science medicine with a psychological and sociological approach to the "whole person" (Lawrence and Weisz 1998). In other respects, they rather complemented each other: Whereas Katsch was a straightforward pragmatist of productivity who focused on a single chronic disease, Weizsäcker was a generalist and a theorist who embraced the interventionist stance of a (self-proclaimed) political adviser and public intellectual.

Productivity and the Pathologies of the Welfare State

Of course, working ability had always been an angle of medical treatment, but in the Weimar *Sozialstaat* (social state) work became a raison d'être. The Weimar constitution moved labor to the very core of the social contract. Article 163 stated that everybody had the right to earn a living through work and, hence, to be compensated in times of unemployment or when unable to work because of ill health. This entitlement was complemented by the obligation to keep oneself healthy.

This concept of citizenship by labor and health rested on a three-pillar system of social security. The first pillar, social insurance, was based on individual entitlements resulting from monetary contributions. As only paid work generated contributions to social insurance, class distinctions grew less important, whereas having a job or not became a major distinction (Stolleis 2003, 126). The National Socialists did not fundamentally change social insurance, although they abolished self-government and introduced the Führer principle. Versorgung (compensation), the second pillar, took care of a substantial number of disabled veterans and their families. It was state-financed and based on entitlements. In contrast, Fürsorge (welfare), the third pillar, served as a subsidiary. It was tax funded and means tested, but recipients of benefits now retained their right to vote (Eghigian 2000, 229; Stolleis 2003, 125). The concept of Fürsorge also covered state intervention when addressing "endangered" groups such as newborns, adolescents, or women in childbed. In Garz, as well as in Heidelberg, physicians made a case for a closer connection of medical treatment and Fürsorge. Among other things, their plans for work therapy also included vocational aid. So, in addition to their medical tasks, the physicians engaged in vocational counseling and job placement, negotiating with both former and potential employers of the patients and welfare institutions (Katsch 1930; 1943; Weizsäcker 1939, 52; Weizsäcker [1931] 1986, 122).[8]

Qualifying diabetes treatment as "productive" seems to have been modeled after a technical term of contemporary social policy, *produktive Erwerbslosenfürsorge* (productive unemployment relief), which referred to public relief work, such as building streets and canals, or cultivating badlands. It was known as an instrument of municipal poor relief since the nineteenth century, but after World War I, it was transformed into a mandatory tool of unemployment relief subsidized by the Reich (see Sachße and Tennstedt 1988, 95). This work focused on infrastructural projects that required a large amount of human labor power and were supposed to "convert unproductive means of support into productive work" (Böhm 1929, 64) by offering paid positions to the unemployed.

It seems that, in this context, the word "productive" was used in a rather narrow, technical sense—in a sense, how a "productive cough" or the "productive stage" of tuberculosis are technical terms of medicine. Still, the protagonists felt the need to explain in what way these programs contributed to the positive development of the national economy, especially as they failed to reduce the number of unemployed people significantly (see Sachße and Tennstedt 1988, 96). As one author wrote, the measures were not economically productive by themselves, but only when related to "aggregate economic needs" (Schulz 1928, 30). This is where medical imagery came in: The productivity of "productive" relief was discussed in terms of its preventive and therapeutic effects. Remaining in the workforce would preserve the unemployed person's capability and will to work (Schulz 1928, 21–2, 30, 34; Scholtz and Herrnstadt 1929; see Crew 1998, 191–2).

Social policy followed preventive, curative, and rehabilitative rationales, but "productive relief" also consisted of elements of compulsory labor, because it promised to test the "willingness to work" with a negative result leading to exclusion from public support (see Sachße and Tennstedt 1988, 96; Crew 1998, 188–92; Schmuhl 2003, 128). This two-faced approach also formed the Garz agenda. Based on a catamnestic survey, the first success record of 1932 detailed medical parameters such as the intake of carbohydrates, weight, and insulin for a large number of cases, but also reported the work status with the notations "social +" or "social –" (Lublin and Kroner 1932). Before and after 1933, one objective of productive diabetes relief was to expose "anti-social elements" (61): patients who did not (try to find) work or defied diet prescriptions in a manner threatening to their survival (Banse and Spickernagel 1939, 28, 34, 42–3; Katsch 1940, 72–3). In addition, in Heidelberg, physicians painted a picture of all sorts of nonworking individuals as (potentially) productive relative to a small minority insusceptible to therapy and guidance (Weizsäcker 1930, 92; Hollmann 1935, 314–15).

The rhetoric of "hopeless cases" employed to characterize these patients allowed for a quite inclusive understanding of productivity. This perspective conformed with a contemporary handbook definition of productivity, maintaining that all kinds of work tasks were productive, so long as there was a demand for them. Only inoccupation was unproductive (Diehl 1932, 260–1). So when the Garz physicians spoke of "productive work" that diabetics could do, this was actually tautological. They seem to have used the adjective "productive" as a catchword that signaled the transformation of money-consuming invalids into people contributing to the national economy in some way or another. At the same time, the ability to be productive extended beyond the realm of wage-earning labor. In the rationale of work therapy, the "reproductive" activities of housewives, school attendance, and apprenticeship equaled work.[9]

Yet how exactly did this encompassing notion of productive work relate to concepts of productivity in the contemporary economic science as indicated in the handbook cited earlier? First of all, one must take into account the perseverance of state-related economics in the German-speaking countries that had started as sciences of state in the second half of the eighteenth century (see Tribe 1988). Similarly, physicians involved in twentieth-century "social medicine" framed their activities with respect to the polity. Beyond this shared state-orientation of economic theorists and physicians, one can expect more direct reverberations of different currents of economic thought with regard to medical work therapy. The earlier cited wording of "aggregate needs" and the call for counterbalancing state intervention in times of crisis refer to the national economy as a governable entity. Thus, they can be seen against the backdrop of emerging macroeconomics (see Tooze 2001, especially 77–8, 143–4, 170; Mitchell 2005). A further context might have been the so-called younger Historical School of Political Economy. In the second half of the nineteenth century, German political economy was a heterogeneous but comprehensive discipline that included both emerging social and political sciences. It claimed to be part of the humanities and contested universal "laws" in economics (see Grimmer-Solem 2003). These political economists were leading figures in the middle-class social reform movement, which turned out to be a forerunner for many of the features of state-driven, interwar social policy (see Sachße and Tennstedt 1988).

While 1920s German political economy disintegrated into different "systems" of thought overwhelmed by dramatic socioeconomic changes (see Köster 2011), its sociological orientation remained alive. The resurgent theorizing of "productivity" frequently referred to Friedrich List's 1841 critique of classical political economy in which he contended that the producers of goods were not the most productive, but rather were those professionals—doctors, teachers, judges, and administrators—who "produced productive forces" (List [1841] 1910, 232–3). Following him and other nineteenth-century writers, economic theorists distinguished "technical" or "commercial" from "national economical" productivity, the latter including anything that was supposed to satisfy the common needs and contribute to the continuance and welfare of society (Spann 1923, 220–1; Diehl 1932, 261).

Such wholesale definitions conceived of economic activities as contributing to social productivity. Thus, its concept was quite in contrast with productivity analysis that was becoming a main tool of interwar econometrics.

A case in point is economic theorist and social reformer Frieda Wunderlich, who, in 1926 published her monograph *Produktivität* (see Wobbe 1995, 172–3). Her concept of productivity put a strong emphasis on the enablement of production. So, in addition to measurable returns on economic action, her definition of productivity included the potential for keeping up and enhancing production in the future (Wunderlich 1926, 128, 350).[10] Under these premises, she regarded welfare policy as a regular branch of economic policy, which served the improvement of productivity (348). Drawing on Rudolf Goldscheid's concept of "Menschenökonomie"—the necessity to systematically cultivate modern societies' human workforce (see Bröckling 2003)—Wunderlich (1926, 51–2, 304) marked the way in which labor power could be "preserved most thoroughly, developed and enhanced in the best way" as a key problem. In contrast with the corporate production of working selves in the 1950s or the privatizing agenda of 1970s motivation psychology (see Donauer 2015; Held this volume), the interwar social reformer addressed the state when designating responsibility for fostering human productivity. According to Wunderlich, the preservation imperative required society to provide "Fürsorge" regardless of means, because "restored or cultivated working power is the reward that will benefit society later on, comparable only to agricultural melioration that pays off slowly, but in the long run generously" (346).

This general statement, linking medical prevention and curative treatment to the physiocratic "riches of the soil" (see the introduction to this volume), fits exactly with the way that the promoters of work therapy legitimized their cost-intensive activities. Both Katsch (1930; 1942–3; 1940, 73, 77) and Weizsäcker (1930, 67) emphasized long-term cost savings for social insurers and welfare offices. The former also advertised diabetics as valuable workers for National Socialist Menschenökonomie.[11] In so doing, they were able to build on a general, post-World War I willingness to invest in the capability of low performers that flourished again when the National Socialists started to prepare for war.[12] So it was certainly no coincidence that in 1932, Viktor von Weizsäcker published an article in the journal *Soziale Praxis*, which Wunderlich edited (Weizsäcker [1932] 1986; 1949, 87). However, whereas insurance-sponsored "productive relief" was in agreement with Wunderlich's outlook on the democratic social state, there were also resonances on the other side of the political spectrum. Political economist Othmar Spann was a well-known proponent of an authoritarian corporative state in the context of Austro-fascism. As a theorist of "fruitfulness", he propagated a holistic model of the economy in which the productivity of "guiding, intellectual work" could assert a prerogative over the "passive fruitfulness of goods" (Spann 1923, 227). This corresponded with Weizsäcker's self-perception as a "geistiger Arzt" (spiritual-intellectual physician) who made sense of illness and provided medical and personal guidance.

On a practical level, Weizsäcker's interference with "economy" was focused on the problem of good government (see Mitchell 2005). One of the rare occasions where he used the term "productive" in a positive sense referred to the intellectual potential of a working group he had founded, bringing together physicians, economists, and sociologists. This group discussed the flaws of the existing social security system and aimed to elaborate an alternative approach. They received funding from the Emergency Foundation of the German Sciences and the Abraham Lincoln Foundation, a secret branch of the Rockefeller Foundation designed to spread a democratic spirit among young German academics.[13]

In a memo outlining his research agenda, Weizsäcker stated that his reform propositions "remarkably resemble those originated from entirely different contexts", for instance, "the expert assessment by Johannes Popitz on the future fiscal equalization scheme".[14] Weizsäcker (1949, 86) met former Secretary of State Popitz and other fiscal policy makers at Alfred Weber's Institut für Sozial- und Staatswissenschaften (Institute for Social Sciences and Governance) in the late 1920s (see Jansen 1992, 35–42, esp. 36, 41). In his report, Popitz (1932, 249–52) argued that the threefold system of unemployment protection should be replaced with unitary emergency relief, although handled in a less disciplining way than "Fürsorge". This was very much in line with what Weizsäcker and his associates proposed. Their reform vision favored a corporative unitary insurance based on profession, which would provide against all contingencies but only be effective when someone was in actual need. Individual assertion of legal entitlements was to be replaced by "a feeling of security amidst the masses of the people".[15]

However, dismissing Weizsäcker as a conservative adversary of the Weimar welfare state is too simple (see Jansen 1992, 58, 242–3, 259). On the one hand, he was very critical of the liberal "social entitlement state" (Eghigian 2000) because it led, he argued, to insecurity and a lack of social cohesion. It is also true that in 1933, Weizsäcker hoped for a genuine national and social reform of the social security system, although he did not exactly welcome the new regime. On the other hand, he did insist on patients being entitled to medical treatment of neurotic working insufficiency. In the years of crisis, Weizsäcker ([1932] 1986, 127) clearly saw that the politics of emergency decrees uprooted the social contract because they de facto replaced the insurance principle for poor relief.[16] He was very much aware of the fact that social benefits were miserably small and that medical assessments of working ability, shaped by legal distinctions and demands even helped to aggravate the situation (Weizsäcker 1930, 48, 80; 1949, 85): "In the strictest sense of the word, every expert opinion is bound to either preserve or to sacrifice the individual's existential interests" (Weizsäcker 1933; 1569).

Against this backdrop, Weizsäcker ([1932] 1986, 129) could be characterized as an unorthodox conservative who borrowed from different spheres of the political spectrum, e.g., when he criticized that social insurance subjected medicine "to the principle of profitability and productivity of the capitalist economic system". This translated into the composition of his working group. Among its members were several "young National Socialists" who had volunteered for the demanding service at the Neuroses Department (Weizsäcker 1949, 92). To name but one of them, Daniel Hauert had joined the NSdAP in 1929, while he was also a member of Count Ernst Reventlow's dissident völkisch circle. He wrote reviews for Othmar Spann's journal *Ständisches Leben*, whereas his dissertation on the social insurance of peasants was published in the journal *Deutsche Agrarpolitik* edited by leading agrarian politician Walther Darré.[17] These völkisch young men (with anti-Semitism being absent from their writings) took a pedagogical approach to winning people over who could still be valuable members of the "Volksgemeinschaft", including former communists.[18] They were prepared to put the blame on a dysfunctional system of social security instead of targeting individual shortcomings.

However, according to Weizsäcker, the "brightest minds" of the group were to be found on the liberal left. One of them, Georg Böse, was an economist teaching at the Institute of Journalism Studies. In the context of the group, he analyzed how the big state insurers were intertwined with the Reich's financial system. With Böse, Weizsäcker went on excursions to visit industrial plants and to meet trade unionists, as well as entrepreneurs.[19] Another liberal member was Dolf Sternberger, who was to become an eminent political scientist in the Federal Republic. At that time, he was a PhD holder in philosophy and a journalist. He investigated how insurers, insurants, and physicians conceived of social laws.[20] These studies were never published, as the Nazi changeover of power led to Böse's dismissal as well as the breakup of the group.[21]

Beyond Productivism: Conceptualizing Work and Performance

Neither Katsch nor Weizsäcker or any of their colleagues came up with a sophisticated notion of "productivity" from a medical perspective. As for Katsch, it seems that he primarily echoed the semantics of social policy. Nevertheless, these physicians did contribute to processes of social engineering that dealt with the working capacity of individuals. For both men, work therapy had an experimental side that was supposed to reshape the understanding of "work" and "performance"—and, prospectively, society's ways of dealing with unfit individuals.

In 1931, Weizsäcker ([1931] 1986, 114) was invited to give a talk at the Reich Labor Ministry. In order to explain his thoughts on the reform of social insurance, he began with the fundamentals: "Modern medicine does not consider man to be a reservoir of energy or a work-providing machine, but an utterly delicate, psychophysical entity". Here and elsewhere, Weizsäcker argued against the gospel of modern physiology that had investigated patterns of fatigue to understand the working body, especially under conditions of industrial production. Anson Rabinbach (1992, 3) has coined the term "productivism" to denote the belief that human society and nature are linked by the convertibility of all productive forces through the common currency of energy governed by natural laws, quantifiable by means of objective natural science, and endlessly improvable. Now, in a situation of economic entropy that rendered the social security system dysfunctional, Weizsäcker (1930, 36–7; 1931, 97–9) held that the notion of work and working ability needed to be fundamentally revised, which implied an abandoning of the rationale for economic profit.

In the 1920s and 1930s, the nineteenth-century thermodynamic fusion of increased production and liberal social reform translated into politically proliferating productivist utopias of "rationalizing" the body and exploiting the mechanics of the mind (see Rabinbach 1992, 271–2, 278–80; Nolan 1994, ch. 9). The idea that maximum exploitation diminished productivity had been a credo of productivism from the start, and in the interwar years, German labor science and labor medicine still searched for forms of work that would allow for a long time span of performance in relatively good health (see Schottdorf 1995, 90–1; Hau 2008, 393–4). That was what Katsch and Weizsäcker wanted for their patients, too. Yet the latter pointed out that "Taylorism, psycho-technics, and physiology of work" were following the logic of a mechanist "Organökonomie" (organ economy) and, thus, failed to conceptualize the working body adequately (Weizsäcker 1931, 99).[22]

Instead of the motor, the "system" became their dominant metaphor. Weizsäcker (1931, 99) pointed to Friedrich Kraus's 1897 *Die Ermüdung als ein Mass der Constitution* (*Fatigue as a Measure of Physique*), a study that had put working ability "in the bigger framework of the stability of a capable system, not of the organ, but of the organism, not of wins and losses, but of proving itself capable". Kraus had been director of the Second Medical Clinic of the Charité in Berlin. Alongside Ludolf Krehl of the Heidelberg Medical Clinic, Kraus was the founding father of whole-person approaches in German internal medicine between 1900 and 1920. Katsch belonged to the Berlin school of "functional pathology" via Kraus's disciple Gustav von Bergmann; Weizsäcker, who was also a physiologist by training, had worked under Krehl for many years.

The idea of bodily (self-)regulation was, of course, not new at the turn of the century but had coexisted with the more prominent mechanic-energetic notions of the body (see Canguilhem 1979; Tanner 1998). Actually, diabetes had been conceived of as a regulatory disease since Claude Bernard's time. Still, the holistic boom around 1900 created the new leitmotifs of "systems" and "organization" in various fields of knowledge and practice (see Heyck 2014). One manifestation of this trend was the physiology and psychology of "function" in the ailing body. The basic assumption was that a disease or an injury did not just leave the organism in a reduced state of capability, but that a "change of function" took place. The interrelated functional systems of the organism would adapt to the new situation and compensate for some of the malfunction (Weizsäcker 1931, 113). As a result, the organic system acquired a new form of stability. This is why Katsch (1937, 401) called well-adjusted diabetics "bedingt gesund" (conditionally healthy).

Work therapy not only allowed for a more adequate observation of patients than the clinical environment, its findings also raised new questions of fundamental character. Katsch successfully promoted a research institute that was to "investigate limits and possibilities of work therapy and of performance . . . under different conditions and with respect to the different clinical types of diabetes".[23] Weizsäcker also mapped out a research center in "pathology of performance" in order to develop new methods of assessing what an organ function could accomplish in a given task.[24] The center did not materialize, though, before he left Heidelberg for Breslau in 1941, where he headed a large, special military hospital for brain-injured soldiers that functioned as a substitute for the research facility (Weizsäcker [1943] 1986; 1949, 93). Regardless of the different research facilities, Katsch and Weizsäcker shared a basic question: What was a person in reduced health able to do—not in an isolated clinical performance test of some organ function, but under concrete working strains?

Although this kind of research entailed motion studies familiar from labor science, rehabilitating the sick body did not serve the contemporary phantasm of efficiency.[25] Regaining sufficiency was a more moderate goal, and it referred to both "productive" work and the ability to cope with a lack of employment opportunities: the sick man's "task being his work or his unemployment, his social advancement, or to content oneself".[26] Thus, work therapy's outlook differed markedly from medical performance enhancement as envisioned during the Weimar Republic and radicalized under National Socialism (see Höfler-Waag 1994; Schottdorf 1995; Hau 2008; Plesser and Thamer 2012). The patient's "performance optimum" did not compare to an average standard or to peak performances, but was individually "defined by his state of health" (Hebel 1940, 12). Furthermore, against the backdrop of National Socialist "labor service" and forced labor, work therapists pointed out that working ability could not be enforced (Brauch 1938, 101; Weizsäcker 1939, 51).[27]

Producing Productive Bodies

In their objective to recreate a potential for production, Garz, as well as the Heidelberg Department, put a strong emphasis on activating and enabling the patient. In doing so, they addressed, in different ways, individuality as a resource of recovery. Yet how exactly were bodies involved in the production of "productive forces"?

First of all, work therapy did not set out to produce bodies in a universal state of health defined by the absence of illness. At Garz, metabolic adjustment—which included measuring out muscular work by the hour—was directed to the working body instead of the largely inactive hospital body (Katsch 1930; 1939; Brauch 1938, 101–02; Banse and Spickernagel 1939, 33). Likewise, the Heidelberg physicians described a discrepancy between "hospital health" and "workplace health" that had to be resolved by exposing the body to gradually increasing strains—from easy walks to steeply ascending paths, from weeding to shoveling (Hebel 1940, 15).[28] There are no descriptions of what the patients exactly did, as work therapy seems to have consisted of actions with a commonsensical function. At Garz, patients worked from Mondays to Saturdays in the kitchen and yard, cultivated sixty hectares of agriculture, tended the vegetable garden, livestock, and poultry, or kept themselves busy in the workshops (Katsch 1930; 1942; Brauch 1938, 101; Banse 1939). The Heidelberg Work Therapy Department had similar routines, although on a smaller scale and with an emphasis on gymnastics, ball games, and sports.

Second, regaining productivity referred to social capability as well, so working within a "community" was central to both facilities (Weizsäcker 1930, 48, 51, 67). At the homes, twelve to twenty (Heidelberg) and thirty to one hundred (Garz) patients lived together for several weeks. Fellow patients were to exert an educative influence on each other with "successful" patients serving as living examples of achievable betterment. Another side of the community structure was that the doctors spent the days with the group, practicing participant observation in work, sports, walking, and conversation (Weizsäcker 1939, 50–1; Hebel 1940, 23). Third, work therapy had a performative quality, as it was supposed to make patients experience themselves as capable by cautiously testing and expanding their limits of capability (Katsch 1939, 33–4). In this process, quantification played a role, as physicians collated clinical observation with lab results and findings from the electrocardiogram or blood pressure measurement. Yet this was supplementary—what mattered in the end was the patient's subjective well-being and overall capability.[29]

Work therapy made use of the patient's individual resources by integrating his or her self-knowledge into diagnosis, but it also demanded active commitment to the process of treatment and recovery. This is especially true for diabetics suffering from a life-long disease.[30] Before the introduction of insulin, the strictest of dietary self-disciplining had been the only available treatment. With the new therapy combining diet and medication, patients had to perform a range of activities on themselves reliably and on a daily basis, such as the injection of insulin and the chemical examination of urine sugar. In addition to the established dietary practices of measuring and calculating, the insulin regime required extensive self-monitoring and documentation, because physicians relied on patients' recordings to align diet prescriptions and insulin doses.

Thus, Katsch (1937, 400) argued, "that you have to work without dogma, that you have to individualize". Patients learned how to counterbalance the forerunners of hypo- or hyperglycemia and how to adjust diet, medication, and muscular activity in response to metabolic ups and downs during the day and under different stresses. Garz diabetologists liked to point out that an experienced diabetic knew more about his disease than the average practitioner or hospital physician. Given that the patient showed adequate comprehension, self-reliance and deviation from the doctor's prescriptions were downright central to the therapeutic concept, if this served the self-regulation of the organism and the patient was stable and able to work (Banse and Spickernagel 1939, 52, 81, 90–1).

To be sure, this did not mean that patients were handed over to complete self-reliance. A considerable number of patients were "uneinsichtig" (undiscerning) from a medical point of view or simply did not get along with the therapy. Yet "well-adjusted" diabetics were also prone to deviate from the straight and narrow. Hence, physicians and insurers surrounded diabetic individuals with a range of supporting and controlling institutions such as social nurses and consultation centers. In fact, it seems that in the interwar years, there was no new medico-scientific body practice that came without psychology (see Hau 2008, 400–6; Jensen 2011). The general trend toward psychologizing different spheres of life, thus decentering and re-contextualizing the body, was central to "holistic" approaches in medicine as well as to all activities that had to do with patient "guidance".

With respect to diabetes, this two-faced approach toward the patient is nicely captured in the German word "Einstellung" which has a significant double meaning: It refers not only to the technical setup of individual daily treatment but also to the attitude of the patient. According to Katsch (1940, 72–3) there were very many diabetics "whose 'will' for disciplined self-stewardship has been paralyzed because of disappointing experiences, feelings of inferiority, . . . discouragement. Their health conscience is not ultimately defective", but could be restored if the physician showed adequate psychological competence. Instead of overanxious self-observation, optimism and a vigorous can-do attitude were to be mobilized. Katsch (1937, 402; 1939, 33–4; 1940, 72–3) thought that no specific psychotherapeutic technique was needed, but that the doctor's educative influence and a normalizing lifestyle in the community of "fellow sufferers" would be sufficient. Thus, work therapy promised to support the "productive development of personality" (Katsch 1939, 32)—in this respect, very much resembling contemporary practices of vocational counseling (see Saxer 2011).

Weizsäcker, in general, did advocate analytical psychotherapy in nervous and internal diseases. Actually, the Neuroses Department was an attempt to make a treatment available for the average hospital patient that hitherto had been restricted to well-to-do, private patients. However, as he considered work-related inabilities to be more or less induced by socioeconomic conditions, individual psychotherapy seemed inadequate. "Situational therapy" at the Neuroses Department and subsequent work therapy made use of group dynamics, and the doctor was supposed to interact with the "environment" of the patient, such as reliving upsetting situations together or interfering with disturbed social relationships (Weizsäcker 1930, 51; Hauert 1935, 579). Weizsäcker even described treatment at the Work Therapy Department as being rather more psychological than somatic, because the focus lay on overcoming blocks, anxieties, and doubts regarding work. Still, it resembled ergotherapy more than talk therapy, primarily because it used the psychological effects of bodily actions (Weizsäcker 1939, 51). So whereas work therapy in internal medicine clearly put the mind to the fore, it did not foster the disembodiment of "productivity", as can be observed in personnel management at that time (see Donauer 2014).

Conclusion

Both Katsch and Weizsäcker not only advocated for the productive potential of sick and disabled individuals, they also put forward, in different ways, the productivity of illness. Weizsäcker generally considered illness an important site of self-exploration, and in 1940 Katsch (1940, 78) recommend the diabetics' low-fat diet and bodily exercise as a model way of life, suggesting that everybody "should feel threatened by diabetes". This revaluation of physical insufficiency relied on a physiology of function that had firmly dissociated itself from the mechanics of ever-increasing industrial growth and efficiency enhancement. Instead, the two work therapists set out to stabilize living systems that were more or less frail.

This, we might assume, corresponded with their experience of the Weimar polity as a complex "whole" in which the interrelated parts produced menacing side effects among each other when placed under enormous stresses. Physicians were confronted with the problems of the welfare state on a daily basis when participating in the allotment of pensions and other allowances conceived of as "health costs". Merging criteria of modern hospital accounting with a traditional sense of good government work therapy can thus be understood as a distinct model for the "economic" management of the working body in illness and health.

Productivity as accomplished in work therapy was about cultivating potentialities that not only focused on working for a living but also encompassed a commitment to the "community" at large. Therapeutic system maintenance required inducing adjustments on different levels. Borrowing from Katsch's term "elastic diet", one could say that patients were trained to mobilize their adaptive faculties by dint of Fürsorge and psychological guidance. Yet the work therapy plan also required that the patient's environment be "elastic"—physicians, family members, employers, health-care institutions, insurance, and welfare administrations had to reorganize in favor of the sick individual's productive potential. Those patients, however, who failed to eventually meet their responsibilities of assisted self-regulation were categorized as "unproductive", expelled from an otherwise remarkably integrative framework of rehabilitation, and relegated to exclusion or receiving repressive treatment.

Notes

1 This chapter was written in the context of the ERC-funded research group "Papertechnology: How Physicians Know, 1550–1950".
2 Circular Reich Ministry for Science and Education, July 4, 1935, Johannes Stein to the rector of Heidelberg University, August 21, 1935, University Archive Heidelberg (UAH), B-6505.
3 Katsch to the curator of Greifswald University, October 17, 1939, University Archive Greifswald (UAG), K 395, fasc. 1, 161. See also Günter Ewert and Ralf Ewert (2010), 164–67.
4 See correspondence in UAH, B-6505.
5 Conversely, see Viktor von Weizsäcker ([1943] 1986); for the routines of the department, see Karl Hebel (1940, 20–4).
6 Weizsäcker to the rector of Heidelberg University, November 7, 1939, UAH, B-6545.
7 See UAH, Acc 30/01 L-V, "Gartenhaus" and "Wielandheim Schlierbach" files. As the patient records of the neurological department await being organized by call numbers, I refer to the designation of the 170 boxes that hold the files. I'd like to express my gratitude to Deputy Director Sabrina Zinke for making the records accessible.
8 See e.g. UAH, Acc 30/01 L-V, box "1929 1126–1155 Gartenhaus 1–90", no. 12 Christian E.; box "1931 Gutachten + Gartenhaus 1–65", no. 10 Philipp F., no. 22 Adam H., no. 24 Philipp K.; box "1932 Gartenhaus 1–77", no. 72 Johann W.; box "1935 617–697", no. 623 Heinrich N.; no. 679 Heinrich R.
9 Weizsäcker to Baden Ministry of Education, December 5, 1930, Generallandesarchiv Karlsruhe (GLA), 235/30343.
10 On the notions of past and future in mid-twentieth century productivity analysis and its gradual shift to the individual's motivation, see Lukas Held (this volume).
11 Weizsäcker to Baden Ministry of Cultural Affairs and Education, December 13, 1928, GLA, 30343.
12 Regarding these backgrounds, see Hans Joachim Banse and Rudolf Spickernagel (1939, 33), Christoph Sachße and Florian Tennstedt (1988, 117–18, 132), Gertraud Schottdorf (1995, 89–91), Michael Hau (2008, 394, 397), Martin Höfler-Waag (1994, 58–92), and Hans-Walter Schmuhl (2003, 258–61).
13 See correspondence in Bundesarchiv (BArch), R 73/16584. See also Viktor von Weizsäcker (1949, 94–5), Wilhelm Rimpau (2010); for the Abraham Lincoln Stiftung see Malcolm Richardson et al. (2008).
14 Weizsäcker, [Memorandum], April 18, 1932, BArch, R 73/16584.
15 See also Viktor von Weizsäcker (1949, 85; [1932] 1986, 135; [1931] 1986, 117–79); Werner Hollmann to Dolf Sternberger, November 11, 1933, Deutsches Literaturarchiv Marbach (DLA), A: Sternberger, folder 2727; Werner Hollmann 1935; Daniel Hauert 1935.
16 See also UAH, Acc. 30/01 L-V, box "1933 Gartenhaus 5–80", no. 10 Philipp E., letter to AOK Heidelberg saying that the emergency decree of December 8, 1931, triggered an aggravation of E.'s neurosis.
17 UAH, PA 4102, file note January 27, 1936; Weizsäcker to Viktor Schwoerer, July 15, 1933, BArch, R 73/16584; Hauert 1933–34; 1935; reviews of *Sozialrechtliches Jahrbuch in Ständisches Leben*, 1933 and 1934.
18 See, for example, UAH, Acc 30/01 L-V, box "1933 Gartenhaus 5–80", no. 35 Oskar K.; box "1937 864–943", no. 893 Heinrich S.
19 Weizsäcker to Schwoerer, July 15, 1933, BArch, R 73/16584; Viktor von Weizsäcker (1949, 86).
20 Weizsäcker to Notgemeinschaft, July 2, 1932, BArch, R 73/16584; DLA, A: Sternberger/Sozialversicherung, folders 5159–62.

21 See Viktor von Weizsäcker (1949, 95); Weizsäcker to Sternberger, October 13, [1933], DLA, A: Sternberger, folder 3776; Werner Hollmann to Sternberger, November 11, 1933 and June 14, 1976, DLA, A: Sternberger, folder 2727.
22 See also Weizsäcker, Abteilung für Leistungspathologie, November 27, 1935, UAH, B-6505.
23 Katsch to Secretary of Science, Education, and Public Instruction, May 5, 1939, UAG, K 395, fasc. 1, p. 111. For the research done on work therapy, see Fritz Brauch (1933) and Fritz Brauch and Helmut Schulz (1939).
24 Weizsäcker, Abteilung für Leistungspathologie, November 27, 1935, UAH, B-6505; Weizsäcker to Baden Ministry of Cultural Affairs and Education, July 31, 1939, GLA, 30336, p. 2; see also Viktor von Weizsäcker (1930, 37–8) and Fritz Brauch (1933, 866) citing Weizsäcker approvingly.
25 For the "efficiency craze" of the turn of the century, see Nina Mackert (this volume). She shows how the discourse on "eating right" in the US contributed to an image of bodily efficiency that was a core aspect of the Progressive Era regime of productivity.
26 Weizsäcker to Schwoerer, July 15, 1933, BArch, R 73/16584.
27 For an overview of different aspects of work in National Socialism, see Marc Buggeln and Michael Wildt (2014).
28 See also among the handful of surviving patient records from the Work Therapy Department: UAH, Acc 30/01 L-V, box "1937 944–1055 Wielandheim Schlierbach 101–110", no. 105 Karl E., no. 106 Karl G.
29 Katsch (1939, 25); UAH, Acc 30/01 L-V, box "1937 944–1055 Wielandheim Schlierbach 101–110", no. 105 Karl E., entries May 26 to June 8, 1938.
30 Oliver Falk is currently working on a PhD thesis that investigates diabetic self-techniques. His study will make use of the Gerhardt Katsch Papers at Greifswald University archive, which had not been accessible during archival research for this chapter.

Bibliography

Ankele, Monika, and Eva Brinkschulte, eds. 2015. *Arbeitsrhythmus und Anstaltsalltag: Arbeit in der Psychiatrie vom frühen 19: Jahrhundert bis in die NS-Zeit.* Stuttgart: Steiner.

Banse, Hans Joachim. 1939. "Der Zuckerkranke im Berufsleben." *Zeitschrift für ärztliche Fortbildung* 36: 167–70.

——— and Rudolf Spickernagel. 1939. *Leistungsfähigkeit und Arbeitseinsatz des Zuckerkranken: Eine Stellungnahme zum sozialen und arbeitswirtschaftlichen Problem des Diabetes mellitus auf Grund 8jähriger Erfahrungen im "Ersten Deutschen Diabetikerheim" Garz auf Rügen.* Leipzig: Thieme.

Bänziger, Peter-Paul. 2015. "Von der Arbeits- zur Konsumgesellschaft? Kritik eines Leitmotivs der deutschsprachigen Zeitgeschichtsschreibung." *Zeithistorische Forschungen* 12: 11–38.

Böhm, G. 1929. "Arbeitslosenfürsorge, wertschaffende." In *Handwörterbuch der Wohlfahrtspflege*, 2nd rev. Edition, edited by Julia Dünner, 64–65. Berlin: Heymanns.

Brauch, Fritz. 1933. "Erfahrungen bei der Arbeitstherapie beim Diabetes." *Zentralblatt für innere Medizin* 54: 865–78.

———. 1938. "Wichtige Gesichtspunkte für die Behandlung arbeitender Zuckerkranker." *Zeitschrift für ärztliche Fortbildung* 35: 100–03.

——— and Helmut Schulz. 1939. "Zur Arbeitstherapie des Diabetes." *Klinische Wochenschrift* 18: 642–45.

Bröckling, Ulrich. 2003. "Menschenökonomie, Humankapital: Eine Kritik der biopolitischen Ökonomie." *Mittelweg 36* 12: 3–22.

Buggeln, Marc, and Michael Wildt, eds. 2014. *Arbeit im Nationalsozialismus*. München: De Gruyter Oldenburg.

Canguilhem, Georges. 1979. "Die Herausbildung des Konzeptes der biologischen Regulation im 18. und 19. Jahrhundert." In *Wissenschaftsgeschichte und Epistemologie. Gesammelte Aufsätze*, edited by Wolf Lepenies, 89–109. Frankfurt am Main: Suhrkamp.

Crew, David. 1998. *Germans on Welfare. From Weimar to Hitler*. New York and Oxford: Oxford University Press.

Dennig, Helmut, and Hermann Becker-Freyseng. 1938. "Arbeitstherapie auf Inneren Krankenhausabteilungen." *Deutsche Medizinische Wochenschrift* 64: 1487–88.

Diehl, Karl. 1932. "Die Lehre von der Produktivität." In *Die Wirtschaftstheorie der Gegenwart*, vol. 2, edited by Hans Mayer, 258–68. Wien: Springer.

Donauer, Sabine. 2014. "Die Körperlosigkeit produktiver Selbste: arbeitswissenschaftliche Konzepte in Deutschland (20. Jahrhundert)." Paper presented at the conference "Produktive Körper. Aktuelle Forschungen zur Körpergeschichte des Ökonomischen", Basel, 5–7 June.

———. 2015. "Job Satisfaction statt Arbeitszufriedenheit: Gefühlswissen im arbeitswissenschaftlichen Diskurs der siebziger Jahre." In *Zeitgeschichte des Selbst. Therapeutisierung—Politisierung—Emotionalisierung*, edited by Pascal Eitler and Jens Elberfeld, 343–71. Bielefeld: Transcript.

Ebbinghaus, Angelika, and Klaus Dörner, eds. 2001. *Vernichten und Heilen: Der Nürnberger Ärzteprozeß und seine Folgen*. Berlin: Aufbau.

Eghigian, Greg. 2000. *Making Security Social: Disability, Insurance, and the Birth of the Social Entitlement State in Germany*. Ann Arbor: University of Michigan Press.

Ewert, Günter, and Ralf Ewert. 2010. *Gerhardt Katsch: Begründer des ersten deutschen Diabetikerheims und der Diabetikerfürsorge*. Berlin: Pro Business.

Feudtner, Chris. 2003. *Bittersweet. Diabetes, Insulin, and the Transformation of an Illness*. Chapel Hill: University of North Carolina Press.

Gebhardt, Karl. 1937. "Erziehungsfragen im Behandlungsgang versicherungspflichtiger Kranker." *Deutsche Medizinische Wochenschrift* 63: 736–38.

Germann, Urs. 2007. "Arbeit als Medizin. Die 'aktivere Krankenbehandlung' 1930–1960." In *Zwang zur Ordnung. Psychiatrie im Kanton Zürich 1870–1970*, edited by Marietta Meier, Brigitta Bernet, Roswitha Dubach, and Urs Germann, 195–233. Zurich: Chronos.

Grimmer-Solem, Erik. 2003. *The Rise of Historical Economics and Social Reform in Germany, 1864–1894*. Oxford: Clarendon.

Hau, Michael. 2008. "Sports in the Human Economy: 'Leibesübungen', Medicine, Psychology, and Performance Enhancement during the Weimar Republic." *Central European History* 41: 381–412.

Hauert, Dankmar [Daniel]. 1933–34. "Der Bauer und das landwirtschaftliche Unfallversicherungsgesetz. Kritik und Reform der deutschen Sozialversicherung." *Deutsche Agrarpolitik* 2: 410–29.

———. 1935. "Sozialpolitische Bemerkungen zur Frage der Kopfunfälle." *Der Nervenarzt* 8: 576–80.

Hebel, Karl. 1940. *Arbeitstherapeutische Erfahrungen: Eine Studie zur Fragestellung der Leistungspathologie*. Leipzig: Thieme.

Heyck, Hunter. 2014. "The Organizational Revolution and the Human Sciences." *Isis* 105: 1–31.

Höfler-Waag, Martin. 1994. *Die Arbeits- und Leistungsmedizin im Nationalsozialismus von 1939–1945*. Husum: Matthiesen.

Hollmann, Werner. 1935. "Arbeitsfähigkeit und Arbeitsunfähigkeit: Zur Grundlegung einer sozialen Therapie." *Therapie der Gegenwart* 76: 261–64 and 311–15.

Jansen, Christian. 1992. *Professoren und Politik: Politisches Denken und Handeln der Heidelberger Hochschullehrer 1914–1935*. Göttingen: Vandenhoeck & Ruprecht.

Jensen, Uffa. 2011. "Die Konstitution des Selbst durch Beratung und Therapeutisierung: Die Geschichte des Psychowissens im frühen 20. Jahrhundert." In *Das beratene Selbst: Zur Genealogie der Therapeutisierung in den "langen" Siebzigern*, edited by Sabine Maasen, Jens Elberfeld, Pascal Eitler, and Maik Tändler, 37–56. Bielefeld: Transcript.

Katsch, Gerhardt. 1930. "Produktive Fuersorge für Zuckerkranke." *Deutsche Medizinische Wochenschrift* 56: 1939–43.

———. 1937. "Garzer Thesen. Zur Ernährungsführung der Zuckerkranken." *Klinische Wochenschrift* 16: 399–403.

———. 1939. *Die Arbeitstherapie der Zuckerkranken*. Dresden and Leipzig: Steinkopff.

———. 1940. "Erhaltung der Arbeitsfähigkeit der Zuckerkranken." *Verhandlungen der Deutschen Gesellschaft für Innere Medizin* 52: 71–78.

Köster, Roman. 2011. *Die Wissenschaft der Außenseiter. Die Krise der Nationalökonomie in der Weimarer Republik*. Göttingen: Vandenhoeck & Ruprecht.

Lawrence, Christopher, and George Weisz. 1998. "Medical Holism: The Context." In *Greater than the Parts. Holism in Biomedicine, 1920–1950*, edited by Christopher Lawrence and George Weisz, 1–22. New York: Oxford University Press.

List, Friedrich. (1841) 1910. *Das nationale System der Politischen Ökonomie: Ausgabe letzter Hand*. Jena: Fischer.

Lublin, Alfred, and Robert Kroner. 1932. *Produktive Diabetikerfürsorge: Ein Bericht über Ziele und Erfahrungen des ersten deutschen Diabetikerheims in Garz auf Rügen*. Leipzig: Thieme.

Mitchell, Timothy. 2005. "Economists and the Economy in the Twentieth Century." In *The Politics of Method in the Human Sciences: Positivism and its Epistemological Others*, edited by George Steinmetz, 126–41. Durham: Duke University Press.

Nolan, Mary. 1994. *Visions of Modernity: American Business and the Modernization of Germany*. New York: Oxford University Press.

Plesser, Theo, and Hans-Ulrich Thamer, eds. 2012. *Arbeit, Leistung und Ernährung. Vom Kaiser-Wilhelm-Institut für Arbeitsphysiologie in Berlin zum Max-Plank-Institut für molekulare Physiologie und Leibniz Institut für Arbeitsforschung in Dortmund*. Stuttgart: Steiner.

Popitz, Johannes. 1932. *Der künftige Finanzausgleich zwischen Reich, Ländern und Gemeinden. Gutachten, erstattet der Studiengesellschaft für den Finanzausgleich*. Berlin: Liebmann.

Rabinbach, Anson. 1992. *The Human Motor: Energy, Fatigue, and the Origins of Modernity*. Berkeley and Los Angeles: University of California Press.

Richardson, Malcolm, Jürgen Reulecke, and Frank Trommler, eds. 2008. *Weimars transatlantischer Mäzen: Die Lincoln-Stiftung 1927 bis 1934. Ein Versuch demokratischer Elitenförderung in der Weimarer Republik*. Essen: Klartext.

Rimpau, Wilhelm. 2010. "'Der Gesellschaftswert ärztlicher Leistungen'. Viktor von Weizsäckers Arbeitsgemeinschaft zur Reform des Sozialversicherungswesens 1932–1934." *Schriftenreihe der Deutschen Gesellschaft für Geschichte der Nervenheilkunde* 16: 335–52.

Sachße, Christoph, and Florian Tennstedt. 1988. *Geschichte der Armenfürsorge in Deutschland*, vol. 2. Stuttgart: Kohlhammer.

Saxer, Daniela. 2011. "Persönlichkeiten auf dem Prüfstand: Die Produktion von Arbeitssubjekten in der frühen Berufsberatung." *Historische Anthropologie* 19: 354–71.

Schmuhl, Hans-Walter. 2003. *Arbeitsmarktpolitik und Arbeitsverwaltung in Deutschland 1871–2002. Zwischen Fürsorge, Hoheit und Markt.* Nürnberg: Bundesanstalt für Arbeit.

Scholtz, Werner, and Ernst Herrnstadt. 1929. *Handbuch der wertschaffenden Arbeitslosenfürsorge: Notstandsarbeiten, Bau von Landarbeiterwohnungen, Umsiedlung, Frühgemüseanbau.* Berlin: Grüner.

Schottdorf, Gertraud. 1995. *Arbeits- und Leistungsmedizin in der Weimarer Republik.* Husum: Matthiesen.

Schulz, Gerhard Heinrich. 1928. *Die Idee der produktiven Erwerbslosen-Fürsorge in Theorie, Praxis und Politik.* Kellinghusen: Hay.

Siebeck, Richard. 1934. "Aufgaben der Klinischen Medizin in der Gegenwart." *Deutsche Medizinische Wochenschrift* 60: 885–9.

Spann, Othmar. 1923. *Fundament der Volkswirtschaftslehre.* 3rd rev. edition. Jena: Fischer.

Stolleis, Michael. 2003. *Geschichte des Sozialrechts in Deutschland: Ein Grundriß.* Stuttgart: Lucius & Lucius UTB.

Tanner, Jakob. 1998. "'Weisheit des Körpers' und soziale Homöostase: Physiologie und das Konzept der Selbstregulation." In *Physiologie und industrielle Gesellschaft. Studien zur Verwissenschaftlichung des Körpers im 19. und 20. Jahrhundert*, edited by Philipp Sarasin and Jakob Tanner, 129–69. Frankfurt am Main: Suhrkamp.

Tooze, J. Adam. 2001. *Statistics and the German State, 1900–1945. The Making of Modern Economic Knowledge.* Cambridge: Cambridge University Press.

Trachte, Heinrich. 1935. "Neue Errungenschaften auf dem Gebiete des Krankenhauswesens der Städt Berlin." *Zeitschrift für das gesamte Krankenhauswesen* 31: 121–4.

Tribe, Keith. 1988. *Governing Economy. The Reformation of German Economic Discourse 1750–1840.* Cambridge: Cambridge University Press.

Unger, Waldemar. 1931. "Ein Versuch sozialer klinischer Psychotherapie." *Der Nervenarzt* 4: 321–32 and 397–406.

Weizsäcker, Viktor von. 1930. *Soziale Krankheit und soziale Gesundung.* Berlin: Springer.

———. (1931) 1986. "Ärztliche Gedanken zur Versicherungsreform." In *Viktor von. Weizsäcker. Gesammelte Schriften*, vol. 8, edited by Peter Achilles, Dieter Janz, Martin Schrenk, and Carl Friedrich von Weizsäcker, 114–25. Frankfurt am Main: Suhrkamp.

———. 1931. "Über den Begriff der Arbeitsfähigkeit." *Deutsche Medizinische Wochenschrift* 57: 1653–7 and 1696–8.

———. (1932) 1986. "Versicherung oder Sicherung?" In *Viktor von. Weizsäcker. Gesammelte Schriften*, vol. 8, edited by Peter Achilles, Dieter Janz, Martin Schrenk, and Carl Friedrich von Weizsäcker, 126–36. Frankfurt am Main: Suhrkamp.

———. 1933. "Vorlesungen über allgemeine Therapie. VIII. Die soziale Krankheit." *Deutsche Medizinische Wochenschrift* 59: 1567–70.
———. 1939. "Arbeitstherapeutische Abteilung außerhalb der Klinik." *Zeitschrift für psychische Hygiene* 12: 49–52.
———. (1943) 1986. "Arbeitstherapie bei Hirnverletzten." In *Viktor von. Weizsäcker. Gesammelte Schriften*, vol. 8, edited by Peter Achilles, Dieter Janz, Martin Schrenk, and Carl Friedrich von Weizsäcker, 187–220. Frankfurt am Main: Suhrkamp.
———. 1949. *Begegnungen und Entscheidungen*. Stuttgart: Koehler.
Wobbe, Theresa. 1995. *Wahlverwandtschaften: Die Soziologie und die Frauen auf dem Weg zur Wissenschaft*. Frankfurt am Main/New York: Campus.
Wunderlich, Frieda. 1926. *Produktivität*. Jena: Fischer.

9 Tracing the Developmentalist Regime of Productivity

Nation, Urban Space, and Workers' Habitat in Mexico City, 1940s–1970s

Monika Streule

In December 1952, only a few years after the grandiose inauguration of the first workers' housing complex in Mexico City, the project's architect and politically well-connected real estate developer Mario Pani published the report "Los Multifamiliares de Pensiones".[1] In this mostly technical publication, Antonio Acevedo Escobedo, editor of the journal *Arquitectura/México*, wrote an essay on everyday life in the new housing units, "Centro Urbano Presidente Alemán" (CUPA), built exclusively for state employees and named after the then president. Giving an account of some day-to-day anecdotes, Acevedo Escobedo (1952, 39–56) intended to counter the conventional public image of the "iron hand"[2] ruling in such "multifamiliares" as pictured in contemporary newspapers, with the success of and admiration for such an exciting "great human experiment". Acevedo Escobedo quoted the administrator of CUPA speaking to a resident and his son, who were accused by the police of having destroyed flowers in the hallway: "I will not do anything to you, even if I could, for being a blight on foreign property. I will wait until you understand the wrong you have done" (54). After some days, the previously furious father, who first insisted he could do as he pleased with his own property, returned to the administrator in order to apologize. The administrator's refusal of punishment had two main outcomes, as Acevedo Escobedo specified: From now on, the plants in the hallway of the disciplined resident were the nicest ones in the whole unit, and residents took it on themselves to denounce unruly children to the administration, thus making the police redundant.

In the context of a prevalent, urban functionalism, such an episode brings to the fore how a number of post-revolutionary Mexican architects and planners envisioned a healthy, modern, and productive future for the public servants living in modernist housing projects. Corresponding to the ideological legacy of the "4ème Congrès International d'Architecture Moderne (CIAM)" of 1933 and the related manifestos of a functional city,[3] such projects, enthusiastically supported by governors and financial institutions, were not only meant to optimize the economics of form through the "meticulous analysis of function" (Moral 1956, 131), but they were also intended to create social change. Moreover, workers' housing estates have been interpreted as the materialization or spatialization of a novel idea of society (see e.g., Fraser 2000; Makropoulos 2003; Caldeira and Holston 2005; Es et al. 2014). Put briefly, against the background of the unprecedented speed and scope of social transformation in post-revolutionary Mexico, such urban projects served as a technique for imposing discipline, progress, and stability in the consolidating nation-state and its capital, Mexico City.

In a broader Latin American context, a more general discourse on economic development had been unfolding since World War II, which was based on the notion of increased productivity as the best means of raising living standards in the fast-growing Latin American cities (Prebisch 1950). At the same time, the emerging developmentalist paradigm was based on the assumption that any population can become more productive if provided with enough capital, technology, and training (Cooper and Packard 1997). All in all, this regime comprised a specific ensemble of political, societal, and academic discourses, as well as theories and statistical standards.[4] The newly inaugurated housing projects were intrinsically shaped by the interests and power constellations engendered by this specific social, spatial, and historical context.

This chapter discusses to what extent the developmentalist regime, as it materialized in modernist housing projects, can be linked to specific ideas of productivity. In so doing, it draws on three of the perhaps most emblematic projects built between 1947 and 1964: CUPA, constructed between 1947 and 1949; a planned industrial city named "Ciudad Sahagún" (1950–1952) in the metropolitan region of Mexico City; and the "Conjunto Urbano Nonoalco-Tlatelolco" (1960–1964). Although, as Priscilla Connolly (1982, 148) and others rightly point out, the allocation of public funds for housing was practically negligible between 1940 and 1964, producing less than 2 percent of the total housing stock in the Mexican metropolis, the narratives construed around them engender a widely ignored dimension of Mexican modernism. Because these housing units were frequently depicted not only as "social experiments" but also as "living showcases", a historical analysis of their development leads to a better understanding of how specific urban social arrangements were transformed in the decades after World War II.

By emphasizing the spatial and corporeal dimensions of developmentalism, this chapter seeks to provide insights into hitherto obscured yet nevertheless hegemonic urban body politics. It challenges the dominant historical narratives in which, as James Holsten (1989, 10) puts it, "Modernism's relation to history is strangely disembodied". Accordingly, this chapter puts forward the question of how urban society established and transformed bodies in specific economic regimes: What role did the body play within specific concepts of productivity, development, and modernity? By addressing the similarities and differences between concepts that have been framed by developmentalism and those that have been shaped by a productivist perspective in a narrower sense, such a perspective helps to deconstruct and to provincialize narratives of progress, growth, and modernity. Whereas explicit notions of productivity are mostly absent from the revised sources of Mexican housing projects and credit schemes, such as promotional short films, architectural journals, and programmatic reports, the concept played a pivotal role in establishing a developmentalist agenda of international agencies such as the Economic Commission for Latin America (CEPAL in its Spanish and Portuguese acronym).

The first section delineates the regime of developmentalism as being intrinsically related to three key elements: a double imperative of economic growth and urbanization, and a nationalist ideology of modernization. The second section shifts attention to the credit arrangements for affordable housing (see Hyman 2011). A close look at newly installed credit schemes reveals to what extent they were embedded in contemporary social hierarchies based on gender, race, and class, thus fostering material disparities by narrowing access to housing and economic rights. Against this background, the chapter concludes that a major shift in the social production of space and productive bodies occurred in Mexico City between the post-World War II era and the 1970s—a change which can be described as a transition from developmentalism to authoritarianism.

Inhabiting Modernity: Social Reform, Urban Productivity, and *Mexicanidad*

Within the discourse of the Mexican modernist movement in architecture and urbanism, a significant role is assigned to the architect Mario Pani. Known as a prototypical exponent of the so-called Movimiento Moderno, Pani certainly promoted such functional paradigms when he presented his ideas for the multifamiliares. He proposed them as a blueprint of social reform, for instance, by characterizing CUPA, the massive housing complex consisting of over one thousand apartments distributed among twelve buildings, as a "novel, daring, and passionate experiment" that beyond its "architectural, urbanistic, constructionist, or economic success" yields "formidable social and human results" (1952, 57). Pani was evidently part of a transnational expert community that has shaped the developmentalist regime. Born in 1911 in Mexico City to a politically influential family, he grew up mostly in Europe and studied at the Ecole des Beaux-Arts in Paris from 1929 to 1934. A few years after returning to Mexico City, he founded the programmatic journal *Arquitectura/México* in 1938, arguably one of the most influential publications on modernist architecture in Mexico (see Flaherty 2013). Pani's planning ideas and architectural discourse clearly took hold when he founded the National College of Architects in 1946. His efforts culminated in his official designation as the architect to design the first massive workers' housing estate of Mexico City. Commenting on the plans for CUPA, Pani acknowledged the influence of Le Corbusier's formula of modern life and functional urban planning as delineated in his 1933 book *La Ville Radieuse* (see Garay 2010).

Yet, in this prevailing but rather truncated narrative of the "Movimiento Moderno", the specific historical context of the contemporary Latin American debates on workers' housing is obscured in three vital aspects. First, current transatlantic debates in urban planning beyond Le Corbusier deeply influenced the architectural debates of the late 1930s (see e.g., Fraser 2000; Almandoz 2002; Domhardt 2012). Against the background of an actually multivocal and frequently contested formation and institutionalization of the disciplines of architecture and urban planning, any one-dimensional narrative falls short.[5] Max Cetto for instance, a Mexico-based German architect and exponent of the International Style of modernism, critically observed the strong impact of a few powerful architects in Mexico City "who have much success and influence", but practiced a "pseudo-modernism".[6] Despite such fundamental criticisms, however, the controversial layout of modernist-functionalist ideals by Pani and others gained strength, and began to play a dominant role in consolidating the curriculum and profession of Mexican architecture.

Second, the sociohistorical conditions of the time are key for understanding the emergence of specific workers' housing policies in post-revolutionary Mexico City. In the early 1940s, the increased housing shortage pushed unionized urban workers to demand a strong social housing policy (see Instituto del Fondo Nacional de la Vivienda para los Trabajadores 1992). Notwithstanding the constitution of 1917, according to which all employers were obliged to offer their workers "comfortable and hygienic accommodation" (Art. 123, XII), until the late 1930s, hardly any housing was provided. Although several socialist architects postulated the urgent need for decent housing for workers again in 1938,[7] only three minor projects were realized with the support of President Lázaro Cárdenas (governing 1934–1940). Other governmental initiatives toward a mitigation of the housing deficit were twofold. On the one hand, the government supported proletarian neighborhoods by authorizing land market or regularizing land tenure in squatted areas. On the other hand, the state introduced rent control within central urban areas in 1942. Workers' housing projects only complemented these strategies, as these domiciles were eligible exclusively for civil servants. Rather than solving any housing problems, they served as "social showcases", representing an idealized life of modern "bureaucrats".

Third, beyond urban planning, there were various interests of the political elite, which particularly encompassed entrepreneurial benefits: On the one hand, President Miguel Alemán's political regime (governing 1946–1952) promoted urban development in Mexico City by forging alliances with local industrialists; on the other hand, the president himself made a private fortune through some of these urban projects by realizing them on his own properties (see Oles 2013, 316). In this way, he worked in tandem with a number of agencies that sought to transform Mexico City and pushed the emerging developmentalist regime.[8]

Against this sociospatial and historical background, the housing projects can be described as one of the aspects of Latin American developmentalism. In a functionalist vision shared by many modernist Mexican architects, urban development was a key focus in the quest for a social change. By increasingly deducing urbanistic evaluations from statistical data, these experts strongly influenced architectural design and urban planning. However, in order to further analyze the modernist housing projects as an aspect of developmentalism, we must move beyond expert knowledge production and discuss to what extent activation strategies played a primary role as well. Understood as the endeavor to activate and regulate the self-governing capacities of the inhabitants (see Bröckling et al. 2011), such strategies were broadly expressed in the official descriptions of the housing projects. Aiming to overcome global economic disparities, CEPAL director Raúl Prebisch (1950, 58; see also Speich Chassé 2013, 208) argued, for instance, that the "maximum rise in the standard of living depends on productivity", and this, in turn, depended "to a great extent on the most efficient machines". Prebisch further underlined the crucial role of "durable consumer goods". Depicted as "continually improving articles offered through technical progress" (1950, 58), his statement illustrates the productivist connotations that accompanied apartments in modern housing estates.

However, the supporters of the project themselves considered the housing estate to be more than a "durable consumer good": the estate was an ideal of spatialized rules and norms where the exclusively eligible public servants would be positively influenced and formed. An essential part of the production of a "common value system among civil servants" (Schmelzer 2013, 15) was a process of socialization into a modern urban way of life and self-identification as a bureaucrat. However, despite the optimistic aspirations by experts, the prevailing public reception of CUPA in Mexico City ranged from skepticism to rejection. In the episode mentioned earlier, for instance, Acevedo Escobedo (1952, 39–40) seems to have been almost offended by newspaper comments of the allegedly strict management of the housing project:

> Through newsletters and caricatures, it was made known that the administrator would mount a sort of permanent guard, smelling people to see if they came with alcoholic breath; after ten at night, there would be a curfew for residents; if the administrator dislikes the gentlemen Gómez, Bigómez, or Trigómez, the elevator service would be denied for all of them.

Vigorously dissociating the CUPA from such techniques of authoritarian bodily control, the author qualifies the media coverage as ignorant, if not envious, antimodernist exaggerations. Against it, his essay and other corresponding publications provide a range of illustrative examples of how workers' housing complexes were supposed to have a positive impact on residents and to forge new behavioral schemes.[9] Consequently, the potential inhabitants of CUPA were depicted as in severe need of "commodious, happy, and civilized flats", in order to overcome their previously "insalubrious" dwelling situations. Likewise, a 1950 article of *Arquitectura/México* reads:

> Those who inhabit such a *vecindad*,[10] work, harbor hopes, retain the instinct of wanting to live decently. Sure, they could have done this already. But they are overwhelmed with bad habits, taste of crap, the vicious habit of being close to the stench and pong of suds and the putrid smell of stagnant pools.
>
> (Arquitectura/México 1950, 312)

Because the residents themselves were not capable of changing this situation, the authors argued that they needed political guidance, modern housing, and education.

Thus, the discourse of these supporters of CUPA recalls what Barbara Weinstein (2015) has coined the "colonial civilizing mission" rather than distinctly productivist ideas. In this equally modernist perspective, urban sprawl was seen as both a natural process and a global phenomenon, which inevitably lead to an aggravated housing crisis and, consequently, to health and security problems. Although this discourse was framed by developmentalist ideas, it drew no direct link between vast population concentration, increasing industrialization, and national productivity. Rather, it proliferated a culturalistic interpretation of poverty (Lewis [1959] 1973; Caldeira 2008, 53).

The contemporary debates focused especially on boys, as they were the ones to primarily "feel in their body and in their consciousness the consequences of social abandoning in the society they live in" (Comité Administrador del Programa Federal de Construcción de Escuelas 1947). In contrast, a childhood in CUPA would offer a different prospect for a healthy child's body, as a resident cited by Acevedo Escobedo (1952, 48–49) enthusiastically wrote: "Here in our increasingly appreciated and dearest multifamiliar (in our CUPA), healthy children are being forged, which is to say, that good children are formed, smart and clean brains for the Mexico of tomorrow". Accordingly, the photos displayed in this essay, such as a snapshot of children playing in the collective open-air swimming pool, depict particularly white and healthy children enjoying CUPA facilities.

Acevedo Escobedo further elaborated on the positive effects that this new living situation had on the male inhabitants: "They have stopped frequenting the centers of vice: they leave work and go to their collective home, where they have everything. Young people too can right there count on places to meet, in a healthy and pleasant environment, without any need to resort to billiards and to houses of ill repute" (44). Such discursive lines strongly remind one of the late nineteenth-century hygienic and social reform debates in the Mexico City of General Porfirio Díaz. Not surprisingly, modernist architects and planners referred extensively to "contemporary urban ills", legitimizing their interventions by a necessity to provide "penicillin for the city" (Arquitectura/México 1950, 309–12). Thus, representing a temporally and geographically specific perspective on desired urban practices, this narrative of hygienic and social reform pushed for a transformation of urban space by combining older social reformist intentions for discipline with productivist ideas regarding laboring bodies, which were at the core of developmentalism.

Regardless of its representation as a dynamically transforming metropolis, Mexico City was divided after World War II. On the one hand was the regime catering to the aspirations of the privileged, and on the other hand was the reality of daily life on the streets.[11] Thus, in an attempt to resolve the tensions that arose from this cleavage, urban housing was mobilized not only to produce and secure modern "bureaucrats" but also to turn underprivileged workers into good citizens of a modern nation-state. In contrast to the previously discussed housing schemes for state employees, the planned industrial city Ciudad Sahagún is an example of a combination of workplace and housing in a remote region of the metropolitan area that was supposed to serve this objective. In a governmental short film produced by the developer in charge, Constructora Industrial Irolo, this paradigmatic showcase of the early 1950s was propagandistically exploited to produce post-revolutionary imagery of the worker as a progressive, young, modern, and urban male individual who replaced the peasant, depicted as backward, nonmodern, and rural (see also Weinstein 2015).

In one sequence of the film, the male narrator comments on the young worker who is "running breathlessly to where his parents are waiting for him . . . He will provide, as a Mexican of the current home country, what the old Mexicans dreamed of for their children: a more cheerful home" (Hernández Bravo 1952, 06:57). In a dramatic climax, the film shows the family moving into their newly prefabricated home, whereas a bulldozer tears down the old adobe one, destroying not only the house but also a monumental agave. The latter served as a symbol of the former *pulque* economy of this region that was based on the production of fermented agave liquor. This quintessential sequence stages what David Harvey (1989) calls "the urbanization of consciousness". The short film was particularly aimed at producing urbanized bodies:

> The New Revolution was fulfilled in places such as Ciudad Sahagún, emerging from the very Mexican Revolution itself. This is now the home of the man who works, who makes Mexico in the factories, in furnaces, in sweat and effort. This is the home, won with straightforwardness and humility. A holy home, clean, full of surprises and admiration for the woman, who giving birth to a son also gives birth to a new home country.
>
> (Hernández Bravo 1952, 07:41)

This aspect of the arguments used to promote Ciudad Sahagún fits well with the conceptualization of productivity as based on a transformation of the workforce by CEPAL. In its developmentalist discourse, industrial workers were also depicted as urban bodies, by contrasting them against the nonmodern, nonproductive rural bodies of the peasants:

> How are sufficient increases in productivity to be achieved? The experience of recent years is instructive. With some exceptions, the rise in employment necessitated by industrial development was made possible by the use of men whom technical progress had displaced from primary production and other occupations, especially certain comparatively poorly paid types of personal services, and by the employment of women. The industrial employment of the unemployed, or ill-employed, has thus meant a considerable improvement in productivity and, consequently, where other factors have not brought about a general lowering of productive efficiency, a net increase in national income.
>
> (Prebisch 1950, 5)

However, whereas Prebisch suggests an increasing employment of women, the assigned role of the female body in the popular Mexican discourse was seen relative to reproduction. As depicted in another sequence in the aforementioned promotional film, the mother, arriving in the family's new home in Ciudad Sahagún, immediately appreciates the kitchen amenities. In this vein, a female reproductive body is delineated alongside the male productive body. Against the background of nation-state formation, which, next to social reformist urbanism and productivism, it forms the third pillar of Mexican developmentalism: The female body is depicted primarily as a mother's body. It is paradigmatically represented in the well-known "Monumento a la Madre", inaugurated in 1949 in Mexico City.

In its need to integrate indigenous bodies, the Mexican project went beyond other modern nation-building ideologies by producing a specific national identity coined as *Mexicanidad* (Lomnitz-Adler 2001). One of the essential strategies for buttressing such a nationalistic discourse was, again, the production of a specific urban space. Covered extensively in the local and international press, the Conjunto Habitacional Nonoalco-Tlatelolco functioned as a symbol for such an imagined harmonious nexus of Mexicanidad, modernist architecture, and the "Partido Revolucionario Institucional (PRI)"-government.[12]

Nonoalco-Tlatelolco, by far a more ambitious and monumental project than any other Hernández project realized before, was planned for one hundred thousand residents in fifteen thousand apartments distributed over dozens of high- and low-rise buildings. With a tabula rasa attitude, demolishing "90 percent of existing buildings of state and private property", Pani (1960, 194) proposed to regenerate Mexico City with his ultimate "gigantic modernist housing complex". The project was initiated in the early 1960s in order to transform former railway yards located in the equally pre-Columbian and colonial center of Tlatelolco. Until then, the area had served not only as an infrastructural intersection and railway workspace but also as a red-light district, with late night bars and vibrant dance halls within walking distance from downtown.

Supporters of the new urban housing development labeled it nothing less than "the crucible of Mexicanidad" (Banco Nacional Hipotecario Urbano y de Obras Públicas 1964). Similarly, in the Spanish edition of *Life* magazine, a favorable reportage on the "spectacular development of Mexico" was published in 1965. As a "visual proof of and symbol for the spirit of this forward-marching country [pueblo en marcha]" (Life En Español 1965), Nonoalco-Tlatelolco was particularly highlighted and closely linked to notions of progress, development, and economic growth. In the cover picture by Alfred Eisenstaedt, indigenous, colonial, and modern symbols are shown in a single, histrionic, spatial arrangement that was supposed to establish a permanent declaration of the nation-state identity of *Mexicanidad* as well as an imagined cultural continuity. This discourse ostentatiously materialized in the layout of Tlatelolco's main square: the "Plaza de las Tres Culturas" (Plaza of the Three Cultures). Both the magazine and the design of the square reproduced the contemporary discourse on nation building. By describing the juxtaposition of historical spaces where "the row of walkers represents three cultures, the Indian, the white, and the mestizo", the *Life* magazine article qualified the square as "an excellent example of what modern Mexico is" (Life En Español 1965).

Generally, in Tlatelolco, and particularly in its main public space, Mexicanidad was materialized as a specific urban space where disciplinary power was employed to develop national consensus. However, it also represented a heavily contested urban space, perfectly suited to perform regime-critical protests. On October 2, 1968, two years after the housing complex had been completed, the government ordered the police and the army to defeat the student's movement, fearing that the mounting number of protests would jeopardize the Olympic Games scheduled to start on October 12 (see Poniatowska [1971] 2007; Gallo 2010; Huffschmid 2015). Several thousand students had gathered on the plaza to protest against the government's repressive policies, when the police and soldiers started shooting. They killed about three hundred people and injured and humiliated thousands more in the course of the subsequent arrests. In this massacre on the Plaza de las Tres Culturas, the state demonstrated its increasingly repressive orientation in a highly symbolic urban space. The popular imagery of this urban space radically changed from "the solution to many of the cities' ills and as an icon of social progress" (Gallo 2010, 73) into a space of authoritarian power. Instead of being a means of discipline, the place became a symbol of repressive control. At the same time, the traumatic historical moment was symptomatic of deeper transformations that ultimately led to an end to the regime of developmentalism.

Becoming a Subject of Credit

Linked to the regime of developmentalism, novel housing loan and credit instruments were developed. As with the modern housing projects and other efforts, they resulted in a specific bureaucracy of exclusion, because the selection processes focused on civil servants and, thus, engendered a particular subject of credit. The previously discussed production of the "bureaucrat" was substantially supported by these newly installed credit schemes. At the same time, they turned a blind eye to the majority of nonbureaucrats in Mexico City who were urging for housing. By unfolding the shifts in the strategies for housing credits, this chapter demonstrates that these changes were neither inevitable nor obvious and that they frequently had unexpected effects on urban space. Not least of all, the exclusion of the majority of the population from credit schemes contributed to an explosive dynamic of popular urbanization in peripheral areas of Mexico City (see Maffitt 2014; Streule and Schmid 2014).

Architect Mario Pani's uncle, Alberto J. Pani, chair of the Secretariat of Finance and Public Credit, played a leading role in coordinating the drastic financial and political reforms of the post-revolutionary era, part of which were the credit schemes. They were initiated in 1924, with a new general law for credit institutions and the subsequent foundation of the central bank, the Banco de México, in 1925. With a designated interest in urban planning (see Pani 1917), Alberto J. Pani was not only a mentor for his nephew, but generally endorsed modernist architecture (see Díaz Hernández 2014). In this vein, he founded the Banco Nacional Hipotecario Urbano y de Obras Públicas S. A. (BNH)[13] in 1933. It was probably one of the most decisive financial institutions for the development of the Mexican workers' housing sector, next to the public pension fund Pensiones Civiles y de Retiro.[14] Both institutions were created generally to provide mortgage loans and credits. Their particular aim was to "solve the housing problem in Mexico City" (Sánchez B. 1950, 110). They were based on a tripartite system, equally funded by employees, employers, and the federal government. The BNH implemented a twofold strategy in order to foster a particular regime of developmentalism that combined real estate credit business with elaborated scientific urban studies: With publications such as *El Problema de la Habitación en la Ciudad de México* (published in 1952), the BNH produced specific knowledge based on statistical data in order to define a framework for their housing credit schemes.

Against the background of this hegemonic discourse in which potential debtors were seen as caught up "between the urgent necessity and the limited capacity of savings" (Gomez Mayorga 1949, 67), the previously mentioned exclusive credit scheme was outlined. Confident of the integrative forces of the developmentalist regime and keen to build a modern nation, only formal state employees were eligible for housing credits and, thus, had access to an apartment in the new workers' housing blocks. Whereas the Pensiones

Civiles financed the development of workers' housing such as CUPA, the BNH's main concern was to provide the individual real estate credits and the financing of urban and industrial infrastructure. In September 1950, it promoted the idea to provide massively cheap housing credits. It started by inviting "all relevant official and private institutions" to a conference to work out a solution for the housing problem (Sánchez B. 1950, 110). After the conference, a permanent advisory board was established to support the government in the implementation of the decisions made during the conference. Among the members of the board were the then-directors of the Pensiones Civiles and the BNH, urban planner Carlos Contreras, and architect Mario Pani, who also became the board's leaders. They represented not only scientific expertise but also the interests of financial and other economic institutions, such as the Confederación de Cámeras Industriales (111).

In the same year, *Arquitectura/México* published a detailed report on this decisive meeting that let readers discern its purpose. In its effort to approach the housing problem, the conference agreed upon three basic steps: First, they argued for technical studies based on statistical data to better understand the urban problems; second, they recognized an urgent necessity to demolish and reconstruct designated "unhealthy or uninhabitable areas, taking into account the social problems that this action derives"; and, finally, they agreed to "promote and guide the construction of economic accommodation, affordable to private and public workers (obreros y empleados), considering financial criteria" (Sánchez B. 1950, 115). A substantial shift with respect to the former credit policy was realized here, as these recommendations included private and public workers. By the 1950s, both groups of workers were considered as being capable of guaranteeing their loans on account of stable incomes and life insurances. As a positive effect of such a strategy, a deregulation of the rent control regime implemented in 1942 was expected.[15] Consequently, the conference concluded, "the best way to solve the serious problem of workers housing is based on joint action of the State, workers and employers, and special credit institutions" (114).

As a reference, the completed projects by the Pensiones Civiles were explicitly mentioned, thus again putting the production of urban spaces such as CUPA in the foreground. However, despite the prominence of Pani's formula of a multifamiliar, where "working people [gente de trabajo], who aspire to have a better life, and who are in the situation to develop an effort to do so, conveniently live" (Sánchez B. 1950, 115), the invited committee rejected the proposition to standardize large-scale housing construction. Moreover, although they acknowledged the economic efficiency of the multifamiliares model, its social implications were questioned, "because large families can hardly be allocated in collective buildings" (Sánchez B. 1950, 115). Given the fact that, in these times, ordinary households in Mexico City consisted mostly of extended families, this reasoning seemingly was compelling.

As a tangible result of these debates and negotiations, the National Housing Institute (Instituto Nacional de la Vivienda, INV) was founded in 1954.

It promoted affordable private housing credit schemes until 1970. The INV was responsible for the construction of approximately fifteen thousand housing units. Additionally, the credit options were substantially diversified in 1963, when the Banco de México established its new housing credit pool, the Financial Housing Aid Fund (FOVI), in order to fund private mortgages for workers' housing. However, despite the efforts to constantly adapt credit schemes and territorial strategies, the vulnerability of this regime became more and more evident in the late 1960s and early 1970s. For example, in a 1968 interview with Mario Pani, the architect overtly reconsiders his previous schemes for modern housing, envisioning a differentiated strategy:

> Mexico City requires a number of housing units that encamp the arriving peasant and put him into something that is still an intermediate between the hut, where he lives, and the house, where he will live in two or three years. Such a housing unit is the place, where he is accustomed and susceptible to become a subject of credit for building him an apartment. To date, we are missing such an accommodation, in which the individual stays in a kind of quarantine in order to learn to live, and not to . . . start making the kind of filthy huts that are a problem everywhere.
>
> (Arquitectura/México 1968, 59)

Furthermore, in the context of explosive urban growth in Mexico City, a number of urban conflicts and social struggles ultimately contested and confronted the developmentalist regime. They emerged in the context not only of the previously mentioned student movement but also of a much broader urban movement of people in marginalized areas demanding infrastructural services and land titles (see e.g., Maffitt 2014). At the same time, conflicts concerning urban policy, not the least in relation to the planned construction of a metro, revealed antagonistic positions within the PRI. A financial crisis that led to the abandonment of the economic model of industrialization by import substitution further destabilized the developmentalist regime (Streule 2016).

Against this backdrop, Pani's statement echoes an ongoing debate that centered on a diversification of the housing model. This resulted, not least of all, in the development of novel credit facilities and was backed by dominant international housing directives from the World Bank and CEPAL. In a substantial reform of constitutional housing law in 1972, the federal obligation to provide affordable housing for workers was explicitly linked with a finally installed "funding system to provide cheap and sufficient credit" (Sánchez B. 1950, 113), fixing company contributions at 5 percent of their workers' salaries. As such, the funding system was based on individual financial contributions, replacing the general pension fund supplied by the state. Accordingly, the workers' housing schemes were diversified, and new institutions were established: the Housing Fund for State Workers, the Housing Fund for Salaried Workers in the Private Sector, and the Housing Fund for Members of the Armed Forces.

A paradigmatic urban project representing this renovated housing policy was "Unidad Habitacional El Rosario", built in 1972 by an ambitious new generation of architects who had close relations with the political and entrepreneurial elite. Exceeding the previous workers' housing complex scheme of the multifamiliares in scale and diversity of buildings, the Unidad Habitacional El Rosario consisted not only of housing blocks but also of privately owned single-family row houses. As such, this new workers' housing scheme was apt to provide a range of options corresponding to the financial capacities of each individual worker. However, despite the unquestionably more inclusive mortgage schemes, the newly established funds still failed to address the housing problems of the vast majority of Mexico City's inhabitants. Although formal employment was an exception, the housing policy exclusively focused on unionized formal workers and employees and was not apt to satisfy the increasing demand for affordable housing.

Conclusion

This chapter has analyzed the links between emblematic modernist housing projects of the decades after World War II and specific ideas of productivity emerging in the context of the Latin American developmentalism as fostered, for instance, by CEPAL. The specific variant of developmentalism coined "Mexican Miracle" is conventionally presented as closely linked to one-party authoritarianism, a specific economic model of import substitution, and the production of knowledge based on growth statistics (see e.g., Joseph et al. 2001). In this context, workers' housing estates were designed to present to both the rest of the country and the Western world the modern nation that Mexico was about to become. However, rather than solving the housing problems, the estates engendered corruption and the expansion of bureaucracy. Also, they could only be established by obscuring the realities of most of the inhabitants of Mexico City.

Productivity has not been a core concept for the implementation of the housing projects. Rather, in this specific context, Mexican developmentalism turns out to be composed of three main elements. Next to the production of the national identity *Mexicanidad* and to social reformist urbanization, productivism was primarily a subtext, albeit a crucial one. At the same time, the reconstruction of some principal narratives construed around the modernist housing projects not only sheds light on the implementation of a specific urban social arrangement but also depicts these projects as key sites for the production and contestation of bodies and spaces. Thus, the developmentalist regime appears as continuously renegotiated and remade.

In contrast to the massive housing units, the simultaneously installed mortgage policies proved to be remarkably flexible and finally outlasted the regime itself. They did so, particularly, by adapting to the fundamental social critique that has accompanied the selective and exclusive mentality of the "Mexican Miracle". At the same time, the trend toward more inclusive credit schemes echoed the shifting context from populist and disciplinary practices toward authoritarian and repressive ones in the context of workers' housing. Instead of being based on general pension funds, the housing policy moved toward individual mortgages, and the huge social housing projects were abandoned in favor of a new model of small-scale blocks and single-family row houses. In other words, this policy reform illustrates a profound transformation of the old developmentalist regime, replacing the patronizing attitude of economists, engineers, and architects designing large-scale housing projects and credit schemes with an individualizing duty for every employee to increase personal savings so as to allow them to obtain a mortgage.

Notes

1 I would like to thank Isadora Hastings, Rob Sullivan, Fabian Vögeli, Daniel Weiss the editors of this volume, and the anonymous reviewers for their support as well as their input and advice at various stages of this chapter.
2 All translations are by the author.
3 See, for example, the Charta of Athens (Le Corbusier [1943] 1973).
4 For scholars who have suggested to capture this geographically and historically specific imaginary of development as developmentalism, see e.g. Sikkink 1991; Robinson 2006; Caldeira 2008, 51–77; Legg and McFarlane 2008, 6–14; Almandoz 2015; Weinstein 2015.
5 See Ramón Vargas and J. Victor Arias Montes (2010, vol. 2) and, for a broader discussion on modernisms, Valerie Fraser (2000), Mark Crinson (2003), and Susan Friedman (2006).
6 Archives of the Institute for History and Theory of Architecture (gta Archives), Zurich, Switzerland, 42-ST-5–182, Max Cetto to José Luis Sert, January 29, 1948. Max Cetto was an active member of CIAM. With the professional and theoretical background of socialist "Baufunktionalismus" (architectural functionalism), he moved to Mexico City in 1939. In his letter concerning possible members for a CIAM group in Mexico, Cetto writes: "The difficulty which I see, is the selection of the right people to form this group. You certainly have noticed what kind of pseudo-modernism is playing the leading rôle [*sic*] here and the architects who are its apostles have much success and influence. Unfortunately it is not as easy to make a clear cut between them and the ones who work and think in accordance with CIAM, as it was 20 years ago in Europe, where the reactionaries were on one side and we on the other".
7 Their "Manifiesto a la clase trabajadora, Unión de Arquitectos Socialistas" (Manifesto to the Working Class, Union of Socialist Architects) was presented in 1938 in Mexico City at the International Housing and Urban Planning Congress (see Schäfer 1938; Sánchez Rueda 2009).
8 Envisioning a modern city for a new middle class, Mayor Ernesto Uruchurtu played a determining role in this urban restructuration endeavor. He was governing for an extraordinary long time, from 1952 to 1966, and was nicknamed the Iron Governor due to his disciplining urban policy. Moreover, within this powerful alliance between architects, urban planners, and the PRI-government, the construction company Ingenieros Civiles Asociados (ICA) played a key role in the realization of housing projects by providing civil engineering knowledge.

Founded only two weeks before the construction of CUPA started in 1947, ICA became one of the main Mexican construction companies in the course of the following decades (see Streule 2016).

9 For a broader discussion of the relation of governmental techniques and space, see Matthew G. Hannah (1997), Stephen Legg (2005), and Stuart Elden (2007).

10 A vecindad is a Mexican tenement house, usually an old colonial mansion in the city center, inhabited by numerous, low-income families.

11 In films such as Miguel Morayta's *Vagabunda* (1950) or Luis Buñuel's *Los Olvidados* (1952), among many others, the directors prominently choose modern buildings as film location for heartbreaking dramas or comedies, contrasting them critically with impoverished neighborhoods and social injustice.

12 PRI is the acronym for "Partido Revolucionario Institucional" (Institutional Revolutionary Party).

13 Today it is the Banco Nacional de Obras y Servicios Públicos.

14 In 1959, the Dirección de Pensiones Civiles was transformed into the Institute for Social Security and Services for State Workers. In 1955, an additional pension fund for members of the army was constituted the Dirección de Pensiones Militares.

15 First implemented as a temporary solution, rent control in central areas of Mexico City persisted until 1992.

Bibliography

Acevedo Escobedo, Antonio. 1952. "La Vida En El Multifamiliar." *Arquitectura/México* 33: 181–4.

———. 1952. "Un Gran Experimento Humano: La Vida En El Centro Urbano 'Presidente Alemán'." In *Los Multifamiliares de Pensiones*, edited by Mario Pani, 39–56. México, DF: Editorial Arquitectura.

Almandoz, Arturo. 2002. *Planning Latin America's Capital Cities, 1850–1950*. London: Routledge.

———. 2015. *Modernization, Urbanization and Development in Latin America, 1900s–2000s*. New York: Routledge.

Arquitectura/México. 1950. "Penicilina Para La Ciudad." *Arquitectura/México* 30: 309–12.

———. 1968. "Mario Pani." *Arquitectura/México* 100: 57–63.

Banco Nacional Hipotecario Urbano y de Obras Públicas. 1964. *Tlatelolco, Crisol de La Mexicanidad*. México, DF: BNHUOPSA.

Bröckling, Ulrich, Susanne Krasmann, and Thomas Lemke. 2011. "From Foucault's Lectures at the College de France to Studies of Governmentality: An Introduction." In *Governmentality: Current issues and Future Challenges*, edited by Ulrich Bröckling, Susanne Krasmann, and Thomas Lemke, 1–33. New York and London: Routledge.

Caldeira, Teresa. 2008. "From Modernism to Neoliberalism in São Paulo. Reconfiguring the City and Its Citizens." In *Other Cities, Other Worlds. Urban Imaginaries in a Globalizing Age*, edited by Andreas Huyssen, 51–77. Durham, NC: Duke University Press.

——— and James Holston. 2005. "State and Urban Space in Brazil: From Modernist Planning to Democratic Interventions." In *Global Assemblages*, edited by Aihwa Ong and Stephen J. Collier, 393–416. Malden, MA: Blackwell Publishing.

Comité Administrador del Programa Federal de Construcción de Escuelas. 1947. *Construyamos Escuelas*. Agosto, no. 1. México, DF: Editorial Galatea.

Connolly, Priscilla. 1982. "Uncontrolled Settlements and Self-Build. What Kind of Solution? The Mexico City Case." In *Self-Help Housing: A Critique*, edited by Peter M. Ward, 141–74. London and Bronx, NY: Mansell.

Cooper, Frederick, and Randall M. Packard, eds. 1997. *International Development and the Social Sciences: Essays on the History and Politics of Knowledge*. Berkeley: University of California Press.

Crinson, Mark. 2003. *Modern Architecture and the End of Empire: British Art and Visual Culture Since 1750, New Readings*. Aldershot: Ashgate.

Díaz Hernández, Lourdes. 2014. *Alberto J. Pani: Un Promotor de La Arquitectura En México*. México, DF: Facultad de Arquitectura UNAM.

Domhardt, Konstanze Sylva. 2012. *The Heart of the City: Die Stadt in den transatlantischen Debatten der CIAM 1933–1951*. Zurich: gta Verlag.

Elden, Stuart. 2007. "Governmentality, Calculation, Territory." *Environment and Planning D: Society and Space* 25 (3): 562–80.

Es, Evelien van, Gregor Harbusch, Bruno Maurer, Muriel Pérez, Kees Sommer, and Daniel Weiss, eds. 2014. *Atlas of the Functional City: CIAM 4 and Comparative Urban Analysis*. Bussum: Thoth Publishers; gta Verlag.

Flaherty, George F. 2013. "Mario Pani's Hospitality. Latin America through Arquitectura/México." In *Latin American Modern Architectures: Ambiguous Territories*, edited by Patricio Del Real and Helen Gyger, 251–69. New York and London: Routledge.

Fraser, Valerie. 2000. *Building the New World: Studies in the Modern Architecture of Latin America, 1930–1960*. London and New York: Verso.

Friedman, Susan. 2006. "Periodizing Modernism: Postcolonial Modernities and the Space/Time Borders of Modernist Studies." *Modernism/modernity* 13 (3): 425–3.

Gallo, Rubén. 2010. "Tlatelolco. Mexico City's Urban Dystopia." In *Noir Urbanisms: Dystopic Images of the Modern City*, edited by Gyan Prakash, 53–74. Princeton: Princeton University Press.

Garay, Graciela de. 2010. "Recordando El Futuro de La Ciudad de México: Testimonios Orales de Sus Arquitectos, 1940–1990." *Alteridades* 20 (39): 11–28.

Gomez Mayorga, Mauricio. 1949. "El Problema de La Habitación En México: Realidad de Su Solución. Una Conversación Con El Arquitecto Mario Pani." *Arquitectura/México* 27: 67–74.

Hannah, Matthew G. 1997. "Space and the Structuring of Disciplinary Power: An Interpretive Review." *Geografiska Annaler. Series B, Human Geography* 79 (3): 171–80.

Hernández Bravo, Fernando. 1952. "Ciudad Sahagún. Un Ejemplo Nacional." Accessed November 30, 2015. http://unhidalgo.com/2009/12/documental-cd-sahagun.

Harvey, David. 1989. *The Urban Experience*. Oxford: Blackwell.

Holston, James. 1989. *The Modernist City. An Anthropological Critique of Brasília*. Chicago: University of Chicago Press.

Huffschmid, Anne. 2015. *Risse im Raum: Erinnerung, Gewalt und städtisches Leben in Lateinamerika*. Wiesbaden: Springer VS.

Hyman, Louis. 2011. *Debtor Nation: The History of America in Red Ink*. Princeton: Princeton University Press.

Instituto del Fondo Nacional de la Vivienda para los Trabajadores, ed. 1992. *Apuntes Para La Historia de La Vivienda Obrera En México*. México, DF: INFONAVIT.

Joseph, Gilbert, Anne Rubenstein, and Eric Zolov, eds. 2001. *Fragments of a Golden Age: The Politics of Culture in Mexico Since 1940*. Durham: Duke University Press.

Le Corbusier. (1933) 1935. *La Ville Radieuse: Eléments d'une doctrine d'urbanisme pour l'équipement de la civilisation machiniste*. Boulogne: Editions de l'Architecture d'aujourd'hui.
———. (1943) 1973. *The Athens Charter: La Charte d'Athenes*. New York: Grossman Publishers.
Legg, Stephen. 2005. "Foucault's Population Geographies: Classifications, Biopolitics and Governmental Spaces." *Population, Space and Place* 11 (3): 137–56.
——— and Colin McFarlane. 2008. "Ordinary Urban Spaces: Between Postcolonialism and Development." *Environment and Planning A* 40 (1): 6–14.
Lewis, Oscar. (1959) 1973. *Five Families: Mexican Case Studies in the Culture of Poverty*. New York: Basic Books.
Life En Español. 1965. "Tlatelolco." *Life En Español*, September 27.
Lomnitz-Adler, Claudio. 2001. *Deep Mexico, Silent Mexico: An Anthropology of Nationalism*. Public Worlds, vol. 9. Minneapolis: University of Minnesota Press.
Maffitt, Kenneth F. 2014. "Nueva Política Social, Viejo Contrato Social: Políticas De Vivienda Y Protesta Urbana En La Periferia De La Ciudad De México, 1960s–1980s." *Historia* 47 (1): 113–32.
Makropoulos, Michael. 2003. "Vergesellschaftung durch Architektur. Gesellschaftstheoretische Aspekte der Funktionellen Stadt." In *Kunst, Macht und Institution. Studien zur Philosophischen Anthropologie, Soziologischen Theorie und Kultursoziologie der Moderne*, edited by Joachim Fischer, and Hans Joas, 577–86. Frankfurt am Main: Campus.
Moral, Enrique del. 1956. "Villagrán García y La Evolución de Nuestra Arquitectura." *Arquitectura/México* 55: 131–2.
Oles, James. 2013. *Art and Architecture in Mexico*. London: Thames & Hudson.
Pani, Alberto J. 1917. *Hygiene in Mexico: A Study of Sanitary and Educational Problems*. New York and London: G.P. Putnam's Sons.
Pani, Mario, ed. 1952. *Los Multifamiliares de Pensiones*. México, DF: Editorial Arquitectura.
———. 1960. "Conjunto Urbano Nonoalco-Tlaltelolco. Regeneración Urbanística de La Ciudad de México." *Arquitectura/México* 72: 183–224.
Poniatowska, Elena. (1971) 2007. *La Noche de Tlatelolco: Testimonios de Historia Oral*. México, DF: Ediciones Era.
Prebisch, Raúl. 1950. *The Economic Development of Latin America and Its Principal Problems*, edited by the United Nations Department of Economic Affairs. Lake Success, NY: United Nations Publication.
Robinson, Jennifer. 2006. *Ordinary Cities: Between Modernity and Development*. London and New York: Routledge.
Sánchez, B. Felix. 1950. "Las Conferencias de Mesa Redonda Sobre El Problema de La Habitación Popular: Organizadas Por El Banco Nacional Hipotecario Urbano Y de Obras Públicas, S. A., Bajo Los Auspicios Del Señor Presidente de La República." *Arquitectura/México* 32: 110–16.
Sánchez Rueda, Guillermo. 2009. "Origen y desarrollo de la supermanzana y del multifamiliar en la Ciudad de México." *Ciudades* 12: 143–70.
Schäfer, Paula, ed. 1938. *Housing in Tropical and Sub-Tropical Countries: International Housing and Town Planning Congress in Mexico*. Bruxelles: Féderation Internationale de l'Habitation et de l'Urbanisme.
Schmelzer, Matthias. 2013. "The Hegemony of Growth: The Making and Remaking of the Economic Growth Paradigm and the OECD, 1948–2010." PhD diss., European University Viadrina Frankfurt (Oder).

Sikkink, Kathryn. 1991. *Ideas and Institutions: Developmentalism in Brazil and Argentina*. Ithaca: Cornell University Press.

Speich Chassé, Daniel. 2013. *Die Erfindung des Bruttosozialprodukts: Globale Ungleichheit in der Wissensgeschichte der Ökonomie*. Göttingen: Vandenhock & Ruprecht.

Streule, Monika. 2016. "Ethnography of Urban Territories. A Transdisciplinary Analysis of Metropolitan Urbanization Processes in Mexico City." PhD diss., ETH Zurich.

——— and Christian Schmid. 2014. "Popular Urbanization: A Comparative Concept in Urban Studies." In *Yearbook ETH Zürich*. Zurich: gta Verlag.

Vargas, Ramón, and J. Victor Arias Montes, eds. 2010. *Ideario de Los Arquitectos Mexicanos: Los Olvidados*, vol. 2. México, DF: Instituto Nacional de Bellas Artes y Literatura.

Weinstein, Barbara. 2015. *The Color of Modernity: São Paulo and the Making of Race and Nation in Brazil*. Durham and London: Duke University Press.

10 Waste or Motivation?

The Productivity Discourse between Past and Future in the Second Half of the Twentieth Century

Lukas Held

Erich Kothe (1950, 2–3; emphasis added), director of the Association of German Engineers explained in a talk on "Die menschlichen Grundlagen der Refa-Arbeit" (On the Human Foundations of Work Time Studies) given at a meeting of the Employers' Organization of North Rhine-Westphalia in Düsseldorf in 1949 what he thought to be most crucial to spur economic productivity:

> We need to *monitor* the working attitude of people; we need to *adjust* production facilities, we need to provide industrial safety, we need to select workers so that they work cooperatively in groups, we need to give *clear orders*; we need to *remove disruptions and blockings*. Then, work will become easy![1]

Ten years later, the emphasis on monitoring, adjustment of production facilities, clear orders, and the removal of disrupting obstacles was understood to be ill advised, because any effort to increase productivity that did not take into account the mental side of the employee was now seen to be incomplete, even wrong.

> We won't motivate through more control or even coercion, not through financial incentives or through praise and appreciation. But, rather, we need to keep in mind that achievement motivation differs from individual to individual and even changes. Therefore, we need to study the individual . . . The master and the engineer of today need to get equipped with a well-grounded knowledge of the human factor, according to which he must act.
>
> (Wiedemann 1960, 162)

This shift indicates a crucial change in how productivity was conceptualized. Thus, instead of pursuing the question "what is productivity", it might make more sense to put the cart before the horse and ask: Productivity to what end? In other words, the slight unease that sets in when reading attempts to produce a final definition of the concept[2] from the historian's perspective arises precisely because of their definite character, because answers to the question of what should be understood by productivity often tend to present themselves as natural laws. Yet as historians, we are supposed to be exploring change. Therefore, I want to ask: *To what end* are work processes evaluated with regard to their productivity? What purposes are served by indicating what one refers to as productivity? To whom is knowledge about specific productivity directed and what can be done with such knowledge? Thus, by examining the use of knowledge regarding productivity, we should be able to trace our way back to the meaning of productivity, and in this case, with regard to the circumstances during the period roughly from the 1950s to the 1980s.

The following will attempt to demonstrate that, starting from approximately the 1960s and appearing more pronouncedly in the mid-1970s, the focus in trade journals on what was understood as productivity in the context of plant and business management shifted from the past to the future. This changing focus was accompanied by a transformation in the understanding of the body. Productivity in the sense of balance sheet analyses was recast into productivity in the sense of potentialities. Taking the place of the retrospective correction of work processes that endeavored to save energy and resources (namely, productivity as balance sheet analysis) were the imperatives of readiness and preparation (namely, productivity as potentiality). The chief concern no longer involved searching for production-related and organizational inefficiencies, wastefulness, and unused reserves, although, admittedly, such scrutinizing continued to occur. Rather, in light of what felt like an increasingly unpredictable and ever-more complicated world, the potential to be prepared for future challenges became more and more crucial. The state of such general readiness ultimately sought to achieve permanent motivation, a commitment that could always be further increased. The decisive objective that one strove for in the name of productivity was the knowledge that one was prepared.

Consequently, research into the psychology of motivation played a major role in changing the meaning of the concept of productivity, not the least of which because the publications of some of its most prominent proponents—David McClelland at Harvard University and Heinz Heckhausen at the University of Bochum—extended their authors' influence far beyond the boundaries of achievement motivation research. According to McClelland's thesis, both the individual employee's professional success and the growth of an entire national economy could ultimately be traced back to the individual's motivation to achieve. As a result of the promise of being able to cultivate and even strengthen this decisive productivity factor through training, economic development became a psychotechnical issue.

For the body, the rediscovery of the will in the guise of motivation meant that it gained a new complementing power. As shall be seen forthwith, motivation was understood as a mechanism consisting of internal human and external environmental stimuli that needed only to be properly adjusted so as to release an energy flow that could propel fatiguing physical capabilities beyond their limits. Being motivated meant not allowing one's fatiguing body to thwart one's momentum. Whereas psychologists such as Heinrich Düker and Karl Mierke still viewed the will as a psychophysical power that, circulating between muscles and mind, could be completely exhausted, David McClelland's and Heinz Heckhausen's concept of motivation conformed to a constellation of stimuli, whose forward-driving vortex did not deplete itself on its own accord. Rather, by "installing" additional stimuli, this vortex could even be strengthened. Whereas this did not overcome the body's fatigue, according to the thesis presented here, this fatigue underwent a decisive reinterpretation, becoming a transitional stage and no longer constituting an ultimate limit. In addition, insofar as motivation—as a new power capable of driving subordinated physical capabilities beyond their limits—could itself be regulated and strengthened through psychotechnical training programs, motivation psychologists were tapping into productivity's quintessential source.

For primary source material, I am using three specialized journals that are sensitive to socio-technological changes, particularly in the 1970s, and that negotiate the question as to how best to prepare workers for the expected but not predictable challenges of the future. The first publication is *Rationalisierung*, a monthly journal published between 1950 and 1982 by the Board of Trustees for the Rationalization of the German Economy (Rationalisierungskuratorium der deutschen Wirtschaft). The second is *REFA-Nachrichten*, the main professional journal for work-time studies of the Association for Work Studies (Verband für Arbeitsstudien), which was reestablished in 1946 (previously, it was the Reichsausschuss für Arbeitszeitermittlung—REFA, first founded in 1921). Lastly, the third source is the internationally oriented *Productivity Measurement Review*, published between 1955 and 1965 by the European Productivity Agency, a semiautonomous organization within the framework of the Organization for European Economic Co-operation (OEEC), which was responsible for the distribution of the Marshall Plan funding.

Searching for Wastefulness, Obstacles, and Reserves

In light of the general shortages (especially of raw materials, labor, and machines), deliberations regarding productivity in occupied Germany during the immediate postwar period were especially topical. To be sure, the debate about operational rationalization measures reached much farther back, at least as far as the German publication in 1923 of Henry Ford's autobiography *My Life and Work*. Nonetheless, the *conceptual* understanding of the term "productivity"—its precise definition and mathematical operationalizability—was not clear.[3] What seemed to be indisputable, however, was that increasing productivity meant procedures whose application promised to make more from less by revealing wastefulness, obstacles, and unused reserves and imposing a regime of frugality (Brodmeier 1950, 97). Particularly, given the limited capabilities of undernourished workers, it was necessary to economize more effectively. Appearing in 1948, the first issue of the *REFA-Nachrichten* noted:

> Since, in light of our situation, we can no longer allow ourselves to waste our labor and raw materials . . . we must set about to increase our performance again systematically. In . . . doing [so], however, we need, more than before, to be conscious of straining the physical and mental powers of the working people as little as possible.
>
> (Böhrs 1948, 1)

Two years later, an article maintained, "Thus, it is necessary to clear away everything that requires unnecessary exertion" (Kothe 1950, 2). That literally meant, first of all, bringing to light wastage and obstacles in the workflow. A new type of visualization, observation, and monitoring was to be introduced into operations and factories:

> Especially significant is an overseeable spatial partitioning. Idling, insufficient exploitation of the machine installations, squandering or loss of material, and also the reduction of work performance could be the consequences of inadequate oversight. To restrict this evil to a minimum, one has increasingly made use of the opportunity to install . . . standardized glass walls. The advantages are bright rooms, undisturbed work, great overview, easy monitoring, etc.
>
> (Rationalisierung 1950, 99)

Along with transparency and more light, however, visualization primarily meant the systematic temporal measurement of workflows, simplifying and standardizing them, and making them more ergonomic (see Saldern and Hachtmann 2009). These oft-described rationalization measures will not be discussed here in detail. However, it must be emphasized that in the early 1950s, discourses about productivity still focused on workers and their limited capabilities as the most important productivity factor.

Heike Knortz has traced the vagueness of the concept of productivity back to the weak position of engineers engaged in practical work, as opposed to the historical school of national economics in the nineteenth century. The national economists rejected the use of universal (and therefore ahistorical) economic theorems as instruments for economic analysis—especially rational quantitative measuring methods—in favor of a historical understanding of the economy that emphasized community. In so doing, they substantially ensured that productivity was understood exclusively in terms of workers' willingness to work, whereas the systematic inclusion of nonhuman productivity factors remained largely absent. Early business management in the German-speaking sphere, therefore, largely ignored scientific management because of the influence of the historical school of economics, as well as its strict demarcation from the practical work of engineers (see Reuss 1960, 26; Knortz 2010, 34, 50; Köster 2011, 36–9; see also Uhl 2014, 139).

This changed in the 1950s. The need for an effective distribution and deployment of Marshall Plan funds led to the founding in 1953 of the European Productivity Agency. In its professional journal, *Productivity Measurement Review*, economists, engineers, and mathematicians struggled to achieve a standardized concept of productivity that would facilitate cross-national comparisons. For the first time, attention increasingly focused on other factors besides workers—namely, the efficiency of machinery, the fluctuating prices and quality of raw materials, and the recyclability of waste.

Now, the difficulty consisted of determining the effective share that each respective production factor contributed to the output. The *Review* pointed out: "How much of the increase is due to the effort of the worker and how much to increased capital investment in improved equipment, buildings and organization, is a question of considerable significance" (Productivity Measurement Review 1962, 30). Closely linked to this consideration was the controversial question of remuneration, as employee representatives could not include increased corporate profits that did not demonstrably result from labor in their arguments for higher wage demands. The more individual factors were to be considered with regard to their contribution to the total product, the more comprehensively an operation had to be observed, documented, and evaluated. This posed questions as to how to establish a standardized recording system and what units of measurement should be used to quantify and record the relevant factors, thereby enabling comparisons with other operations. Yet, apart from such questions, the multifactor approach tended to turn into an overly excessive productivity analysis, because each production element, no matter how marginal, could potentially prove to be critical in the final balance sheet analysis and, therefore, needed to be taken into account.

In light of this tendency, a contributor to the *Productivity Measurement Review* felt the need to note "the value of the information gained by such studies should be higher or at least equal to the expenses spent on them" (Knayer 1955, 4). Accordingly, a productivity analysis was profitable only if rectifications of the production process actually produced savings, and that could only be the case if significant wastage or reserves had previously been revealed. Moreover, the complex interplay between the various production factors meant that savings in one area (for example, the recycling of waste) could be negated elsewhere (for example, higher prices for raw materials). Hence, adjustments and reorganizations in the production process were not actually oriented toward a stable ideal but rather constituted a constantly recurring and never fully completed process: "there is always some degree of imperfection in the operations of a firm" (Richman 1958, 22).

According to this understanding, at its core, productivity analysis was supposed to provide decision-making knowledge—namely, "reliable figures on which *decisions can be made with confidence*" (Productivity Measurement Review 1955b, 3; emphasis added). This knowledge was addressed to "the accountant, to the engineer, to the personnel director, and to the supervisor" (Richman 1958, 22; emphasis added) and provided the basis for planning future actions. This perspective also implied a worldview in which future changes were foreseeable and, above all, rather insignificant, so that one might be permitted to react to them with ex-post adjustments (see Siegenthaler 1993, 84–106, especially 87–8). In other words, productivity in the sense of a balance sheet analysis that, as decision-making knowledge, focuses on the past in order to learn from its "mistakes" presupposed a stable, general framework and manifested itself in rectifications. "Productivity measurement", an article in the *Productivity Measurement Review* summarized, "can illustrate how better performances are possible *under similar conditions*" (Weigel 1961, 56; emphasis added).

After a ten-year run of the *Productivity Measurement Review*, the OECD stopped publishing the journal in 1965, despite the fruitfulness of its specialized discussions. Productivity experts from various countries had developed a range of directly applicable procedures and formulas to determine productivity. However, during a time of growing economic interdependency (e.g., within the European Coal and Steel Community (ECSC) and the European Economic Community (EEC)), the *Productivity Measurement Review's* microeconomic focus no longer seemed appropriate. As was pointed out in the final issue of the *Review* (Productivity Measurement Review 1965, 4) itself,

> when the O.E.C.D. were reviewing their aims and objectives, the bias (of the Review) in the past towards plant level productivity measurement . . . meant that when it was decided to suppress these specifically microeconomic activities, the Review itself was scheduled for discontinuation.

This justification already revealed signs of the shifting focus that the productivity discourse underwent during this period.

The Mechanisms of Motivation

In the 1970s, the previous notions of feasibility gave way to a "deep . . . insecurity about the future" (Nützenadel 2005, 352). The *trente glorieuses* in continental Europe, with their historically exceptional high growth rates, were coming to an end (Piketty 2014, 97). The optimism and expectation of stability, which had imbued this exceptional period, eroded. In addition, structural economic changes led to the emergence in the 1970s of an "industrial service society" (Rödder 2004, 12) where, in light of rapid technological progress, the employee's qualifications and potential for further qualification became increasingly important, whereas the physical capabilities played less and less of a role. It seemed no longer sufficient to await the future's challenges and adjust to them after they manifested themselves. Rather, one needed to be preventively prepared (see Donauer 2013, 354–60).

Outlined here in brief, these changes in the general situation resulted in the renewed focus on the "human factor" (Kocherscheid 1975, 254), yet this time the focus was no longer on the person's physical capabilities and economically effective deployment but rather on the person's mind. This shift needs to be seen in the broader context of the gradual differentiation of human resource management since the 1960s, for which to gain access to the immaterial resources of employees became the primary concern (see Bernet, and Gugerli 2011, 441). Although attempts to appeal to a more collaborative relationship between employees and employer can already be seen in the 1920s, as Karsten Uhl (2014, 347–50) has pointed out, the specific mobilization of motivation was quite new because the concept of motivation itself was new.

In the body of primary sources examined here, the keyword "motivation" has appeared sporadically since the 1960s, reaching its highpoint in the mid-1970s.[4] In a programmatic article, Ernst Wolf Mommsen (1970, 4), the chair of the Board of Trustees for the Rationalization of the German Economy, described the future crux of the matter for increasing productivity as follows:

> To an ever-increasing degree, our society and our economy will have to endeavor to motivate the working person of tomorrow, that is, therefore, the learning person of today. Only through real motivation will the person become not merely an *Arbeitnehmer* [an employee, but literally translated, one who takes work] but rather a *Mitarbeiter* [an employee, but literally translated, one who collaborates]. The real peak performances of the person of tomorrow will have to lie in the area that we define as creativity. Therefore, I would also like to see that motivation and creativity are included as the two central concepts in the permanent education process that is implemented at school, at the university, and on the job.

Thus, the Arbeitnehmer that Mommsen had in mind were rather unmotivated. Transforming them into committed Mitarbeiter meant awakening their creativity and willingness to learn so as to become able to confront the future challenges of a "technologically influenced world" autonomously, rather than passively waiting for orders from superiors. The problem seems to have been, however, that the stimulus for such motivation could no longer lie in the concrete and (ideally) exciting new knowledge content, because this content, of course, remained largely unknown; rather, it had to be located in the "permanent education process". Also, because the mere learning aspect of knowledge acquisition could no longer be the actual source of pleasure and satisfaction, the driving momentum—namely, the motivation to learn and perform—had itself become the object of learning.

Both the ability to mobilize willpower (motivation) and the capacity to bring forth something new (creativity) were formulated here as technologies of the self, skills needed to prepare for future challenges. That motivation and creativity were at the same time understood as purposeful and learnable constitutes the decisive difference from the artistic model, particularly the eighteenth-century concept of genius. Whereas creative genius was something special because it focused all its capabilities on generating a unique work of art, the creative employee resembled his or her colleagues, while the learned strategies for originality were deemed *creative* only so long as they increased profit (see Bröckling 2006).

The operationalization of motivation, in particular, was made possible through psychological research in achievement motivation, which, through its studies, laid the foundation for the new productivity discourse by providing the kind of knowledge employers where seeking. Harvard researcher and social psychologist David McClelland put forth a thesis in his 1961 global study, *The Achieving Society*, which stated that the productivity of a country's national economy depended substantially on the achievement motivation of its population. In 1966, the West German publisher Kohlhammer Verlag released a German translation of the work titled *Die Leistungsgesellschaft: Psychologische Analyse der Voraussetzungen wirtschaftlicher Entwicklung*, and an advertisement for the book appeared in *Der Spiegel* magazine, which later also gave the book a favorable, full-page review (see McClelland 1966; Der Spiegel 1967). According to McClelland's thesis, economic growth was not fundamentally dependent on geographical location, climate, the level of societal development, population growth, or the degree of societal division of labor, not to mention "race" but rather on the degree to which individual achievement motivation was developed within a population. This motivation—and herein lay McClelland's key methodological point—could be observed by means of the achievement imagery in a person's mind. By achievement imagery, McClelland (1961, 40–4) meant, first and foremost, a concern with a "standard of excellence" expressed, for example, in a person's stated desire to succeed, but it was also evident, more generally, in the way people thought about overcoming obstacles.[5] The more people thought about gaining achievement goals and overcoming obstacles, the more they showed themselves to be actually motived for action.

McClelland's study evaluated more than 1,300 imaginative stories from schoolbooks used by second- and fourth-grade children[6] in twenty-three countries during the earlier period, around 1925, and forty countries during the later period, around 1950, using a coding process that made it possible to count the "achievement-related ideas" that appeared in those texts (see McClelland 1961, 71–3).[7] In a second step, for each respective country, McClelland related the frequency of such achievement imagery to the country's cycles of economic growth. In doing so, he identified a clear pattern: For countries that underwent a phase of economic growth between 1920 and 1955, it could be shown that the immediately preceding phase had featured an increase of achievement imagery in popular storytelling—thus a strong motivation for achievement occurred. Conversely, McClelland could also show that a phase of economic decline was preceded by a significant decrease of achievement imagery.[8]

The great model for this observation was Max Weber's thesis on the importance of the Protestant ethic for the spirit of capitalism, which McClelland (1961, 70) explicitly intended to adopt and expand. Because of the extremely broad scope of the underlying data, this bold thesis gained not only credibility but also penetrative power, especially through a series of validating follow-up studies—some of them quite recent (see e.g., Bradburn and Berlew 1961; Mazur and Rosa 1977; Arnolds and Boshoff 2003; Beugelsdijk and Smeets 2008; Engeser et al. 2009). Productivity—seemingly now virtually proved—was a psychological factor. This meant that it should be possible to regulate macroeconomic growth by accessing this factor.

McClelland's German colleague, Heinz Heckhausen, a psychologist conducting research in Bochum, assisted significantly in refining the motivation concept.[9] Through numerous individual experiments in the early 1970s, he came to the conclusion that motivation needed to be understood as a complex interplay of intrinsic dispositions and extrinsic conditions of stimulation, as an interaction between the person and the environment: "Today, one can confidently say that 'motives' in the sense of entities or qualities 'in' the person do not exist" (Heckhausen 1973, 31). Therefore, Heckhausen spoke primarily about a "psychological ecology of motivating and non-motivating stimulus conditions" (33) while, at the same time, distancing himself from older stimulus-response theories, which to him neglected the importance of individual perception and interpretation of the environment according to personal experiences (see Heckhausen 1987; 1976, 138–41).

Frederick Herzberg had already made such a distinction in 1959 in his two-factor theory, but without developing this into his own theory of motivation. He stated openly: "Only the goals can be specified from the results of our study. *The precise way in which increased motivation can be developed must await further research*" (Herzberg 1959, 132; emphasis added). For him, motivation was an effect of largely unspecified self-realization opportunities on the part of workers in their work (the opportunity to realize their own achievements, recognition from superiors, responsibility, potential for development and advancement) and the absence of disruptive factors in the work environment, which, in the sense of good "work hygiene", needed to be removed (see Herzberg 1959, 113–19; 1968, 177–92). His key contribution lay in pointing out the genuine variety of disruptive factors, on the one hand, and motivating factors, on the other.[10] Yet, ultimately, Herzberg was unable to explain how work processes could be specifically designed to engender a motivating effect, although he did already assume that motivation must be something distinct from interest when he pointed out "that the jobs themselves have to be set up in such a way that, *interest or no*, the individual who carries them out can find that their operations lead to increased motivation" (Herzberg 1959, 133–34; emphasis added).[11]

Heckhausen did not draw upon Herzberg's two-factor theory in his approach, because, ultimately, he rejected the notion that motivation was linked to the type of activity or its more or less interesting content.[12] On the basis of an understanding of human beings as instrumentally rational agents, Heckhausen proceeded from five fundamental human concerns, which every person endeavored to satisfy or—viewed obversely—to avoid: the concern for connection (obversely, the fear of rejection); the concern to gain power (obversely, the fear of losing power), the concern for success (obversely, the fear of failure), the concern to payback insults suffered by other people (obversely, the avoidance of aggression toward others), and the concern to help without expecting a compensation (no obverse form given).[13] With regard to achievement motivation, this meant that an action could always be understood as an achievement-motivated action if three factors were given. First, the environment had to address the inner concern for success or the fear of failure in a person (starting situation with stimulus for action). Second, this addressive element had to be described as a realistically achievable goal and actively strived for (action with an expectation of success). Third, the goal's achievement—that is, the accomplishment—could be traced back to one's own effort (attributed to oneself) and at the same time evaluated against a generally binding standard of excellence. To be sure, the individual character of the concern for success or the fear of failure differed from person to person according to his or her personal experiences. Yet the way these basic concerns—under specific situational conditions and with regard to the expectations these conditions awakened—functioned as driving forces could be described schematically.

To the extent that motivation research shifted its focus from internal human factors to the structuring of activities under specific situational circumstances, motivation became an adjustable quantity that lost the characteristic of limitations, something that had distinguished the older concept of will in the tradition of Narziß Ach (see Düker 1931; Mierke 1955). People could be trained to first recognize relevant motivation factors in a given situation and then organize them such that they developed a driving effect. To start, this required introspection in order to achieve clarity about one's own needs, hopes, and anxieties—thus nothing less than finding definitive answers to the questions of who one is and what one wants in life. Independently setting realistic goals in a given stimuli situation, effective planning to achieve these goals, aligning action with binding quality standards and the feedback of others, and the self-attribution of success (and failure) could then be learned as psychological techniques (see e.g., Bossong 1978).

In this way, action would become a process of self-structuring and the environment—especially one's fellow human beings—would become a field of potentially stimulating factors. As long as the balance of affects remained correct, thus, as long as the system generated positive emotions during self-evaluation, this system would continue to feed from itself and generate propulsion. In this sense, action was no longer based on the transformation of inner psychophysical willpower into kinetic energy but rather was derived from a kind of vortex that virtually swept the physical powers along. When psychologists such as Karl Mierke had situated willpower exclusively within human beings, they still assumed that it must be exhaustible. In his study on the ultimate limits of human performance, he wrote, "It is possible to convert only as much potential energy into kinetic [energy] through the will as is actually present in the individual case" (Mierke 1955, 284).[14] However, within the concept of motivation, the "distribution" of this power to external stimulating factors in the sense of self-appointed goals at the same time entailed a departure from ideas of exhaustibility. Motivation could ultimately be understood as strategic knowledge regarding one's own structure of concerns, which made it possible to align the action situation with one's own person such that the situation would draw one forward. Those familiar with the mechanisms of motivation and, thus, capable of aligning the interplay of individual factors to their own benefit, would shift themselves into a forward-driving cycle of self-empowerment that no longer contained any fatiguing forces.[15]

These results found their way into operational management through consultants such as Lutz von Rosenstiel. In his 1972 book, *Motivation im Betrieb* (Motivation in the Company), he synthesized the recent research by leading scholars in the field, such as Heckhausen and McClelland, while simultaneously presenting the material in such a comprehensible manner that employees could learn how to motivate themselves. The author claimed to answer to a growing demand "from staff members in many branches" by promising a way of increasing job performance and employee satisfaction at the same time (see Rosenstiel 1972, 11, 90–8; Oehme 1979; Kramer and Winter 1980).

Whereas within the business world, the older concept of productivity in the sense of a balance sheet analysis was reserved for management positions as decision-making knowledge, the new concept of productivity transcended this sector and directed itself toward individuals even before they entered professional life. In this respect, given an uncertain and unpredictable future, the focus of productivity now lay in the potential for action. This impression is confirmed by the operations consulting literature of the 1980s. Thus, for example, the book *Produktivität und Menschlichkeit: Organisationsentwicklung und ihre Anwendung in der Praxis* (Productivity and Humaneness: Organizational Development and its Application in Practice), published in 1984, states:

> Organization development is a new path to the development of . . . for example, industrial companies . . . with the goal of an active and flexible

> adaptation to the challenges of a constantly changing world . . . Decisive for the planned development process is not the problem in itself—the starting situation in which the organization finds itself—but rather the experience of the organization members, which becomes motivation for change.
>
> (Becker and Langosch 1984, 26)

Motivation for change had become more important than problem analysis. The decisive factor for increased productivity was no longer the search for wastage but, rather, the mobilization of new powers for ever-imminent adaption efforts.

Conclusion

As it has been elucidated up to now, before and after the conceptual shift that occurred in the 1970s, the term productivity in each case referred to a specific knowledge—and this knowledge differed strongly with regard to its purposes, addressees, and understandings of the body. Returning to the central question posed at the outset—productivity for what?—one can state in conclusion: Productivity in the sense of a balance sheet analysis was a quantitatively constituted decision-making knowledge (form of knowledge) that served operational management positions (addressees) so as to rectify work processes (purpose). In accordance with this understanding, like other material resources, the employees' physical power was a resource that needed to be economized (understanding of the body). Productivity in the sense of potentiality was a psychotechnical knowledge of self-regulation (form of knowledge) that was directed toward all employees (addressees) for the purpose of motivational preparedness for future challenges. According to this understanding, physical exhaustibility became a transitional stage, to the extent that this knowledge gained the status of being an additional power that itself no longer admitted to factors of fatigue (understanding of the body).

The two concepts of productivity formed out of specifically diverging worldviews. Rectifications of production on the basis of balance sheet analyses were based on the notion of more or less stable, general framework conditions, hence upon an implicit optimism for the future, as was characteristic of the *trente glorieuses*. The imperative for preparation in the sense of permanent motivation appeared as a response to a crisis-riddled future, whose challenges were admittedly expected but could not be concretely known. Although productivity in the sense of balance sheet analysis was certainly not superseded by productivity in the sense of potentiality as described here, a shift in emphasis on motivational preparedness as an important factor of productivity can be clearly seen.

Translated by Bernard Heise.

Notes

1 I would like to thank the editors of this volume for the invaluable feedback I received as well as the participants in the research seminar of Professor Philipp Sarasin at the Research Center for Social and Economic History at the University of Zurich.
2 For early attempts to define the concept of productivity, see Schwake (1951), Hoth (1958) and Reuss (1960). Regarding the conceptual history, see this volume's introduction, Hentschel (1984), and Knortz (2010).
3 The editors of the *Productivity Measurement Review* lamented in the first issue of 1955, "The concept of productivity itself needs clarification. At present, the most convenient definition and the most relevant contents of the concept seem to differ from one sector or from one level to another" (Productivity Measurement Review 3, 1955a).
4 Also, compare the curve for the keyword "Motivation" in Google Ngram Viewer for the period 1900–2008 in the German corpus.
5 It is important to note that David McClelland (1961, 104–05) highlights a difference between "goal" orientation and "means" orientation. Clearly his understanding of "achievement" should not be confused with "success": "Psychologically speaking . . . *n* Achievement is not only more frequently present in stories from more rapidly developing countries but when it is present, it is more apt to be 'means'-oriented rather than goal-oriented. The achievement sequence more often dwells on obstacles to success and specific means of overcoming them, rather than in the goal itself, the desire for it, and the emotions surrounding attaining or failing to attain it. The adaptive quality of such a concern with means is obvious: a people who *think* in terms of ways of overcoming obstacles would seem more likely to find ways of overcoming them in fact".
6 McClelland took these stories to represent "popular culture" and, therefore, the motives of entire populations. Critics found this methodological approach in a questionable way universalistic, because it presupposed that business behavior would take the same forms in different societies, whereas it applied measures of individual motivation to collectivities (see e.g. Rubin 1961; Katona 1962).
7 McClelland had more than sixty assistants at his disposal for the study, which speaks to the project's financial resources and, thus, the general interest in its results.
8 It is important to note here that the achievement motivation being measured always belonged to the phase *prior* to the subsequent phase of economic growth or decline.
9 Heckhausen established the field of motivation psychology in West Germany. He studied with the Gestalt psychologist Wolfgang Metzger, was fascinated by Kurt Lewin's theory of motivation, but was mostly influenced by the research efforts into motivation undertaken in the US by psychologists such as Henry A. Murray and John William Atkinson.
10 However, it was precisely this strict distinction that later came under criticism. See, for example, Dabscheck's (1994) criticism, occasioned by a new edition of *Motivation to Work* in 1992.
11 See also the attempt by Zink (1975, 477–87), to refine Herzberg's thesis on the basis of expanded data and to make it fruitful for the Federal Republic of Germany.
12 It is worth noting that Heckhausen at no point references Herzberg in his synthesis *Motivation und Handeln. Lehrbuch der Motivationspsychologie* (Heckhausen 1980).
13 Heckhausen does not claim this list of concerns to be exhaustive. However, his theory of motivation is very much founded on the assumption that such

fundamental concerns could be designated and that motives are ultimately rooted in them. He speaks of "concerns", rather than "needs", so as to point to their learned character and, therefore, their individual formation (see Heckhausen 1976, 146–8).

14 Karl Mierke (1896–1971) was a student of Narziß Achs (1871–1946) who, along with Kurt Lewin (1890–1947), was one of the German founders of modern psychology of the will, which was also one point of departure for Heinz Heckhausen.

15 It is important to stress that the inner and outer determinants of action need to be understood in a psychological manner rather than in a quasi-physical manner. That is to say, the interplay between inner-personal and environmental conditions is mediated by individual perception, interpretation, and expectations. Therefore, individual perception, interpretation, and expectations do play an important role in how situational conditions function as action triggering cues. However, it does seem appropriate to describe the interplay of inner-personal and environmental conditions as a kind of mechanism to stress their mutual dependence and influence—that is to say, the ecological character of motivation (see Bolles 1974, 18–19; Heckhausen 1980, 175; Weiner 1980, 440–4).

Bibliography

Arnolds, Cecil A., and Christo Boshoff. 2003. "The Influence of McClelland's Need Satisfaction Theory on Employee Job Performance: A Causal Study." *Journal of African Business* 4 (3): 55–83.

Becker, Horst, and Ingo Langosch. 1984. *Produktivität und Menschlichkeit: Organisationsentwicklung und ihre Anwendung in der Praxis.* Stuttgart: Lucius und Lucius.

Bernet, Brigitta and David Gugerli. 2011. "Sputniks Resonanzen. Der Aufstieg der Humankapitaltheorie im Kalten Krieg – eine Argumentationsskizze." *Historische Anthropologie* 19 (3): 433–446.

Beugelsdijk, Sjoerd, and Roger Smeets. 2008. "Entrepreneurial Culture and Economic Growth: Revisiting McClelland's Thesis." *The American Journal of Economics and Sociology* 67 (5): 915–39.

Böhrs, Hermann. 1948. "Der neue Weg der Refa-Arbeit." *REFA-Nachrichten* 1 (1/2), 1–3.

Bolles, Robert C. 1974. "Cognition and Motivation: Some Historical Trends." In *Cognitive Views of Human Motivation*, edited by Bernard Weiner, 1–20. London: Academic Press.

Bossong, Bernd. 1978. *Motivationsförderung in der Schule.* Weinheim: Beltz.

Bradburn, Norman M., and David E. Berlew. 1961. "Need for Achievement and English Industrial Growth." *Economic Development and Cultural Change* 10 (1): 8–20.

Bröckling, Ulrich. 2006. "On Creativity: A Brainstorming Session." *The Learning Society from the Perspective of Governmentality*, special issue of *Educational Philosophy and Theory* 38 (4): 513–22.

Brodmeier, Herbert. 1950. "Mit weniger Arbeit mehr Erfolg." *Rationalisierung* 1 (4), 97–9.

Dabscheck, Braham. 1994. " 'The Motivation to Work' by Frederick Herzberg, Bernard Mausner, and Barbara Bloch Sneyderman." *Journal of Economic Issues* 28 (1): 297–300.

Donauer, Sabine. 2013. "Emotions at Work—Working on Emotions: The Production of Economic Selves in Twentieth-Century Germany." PhD diss., FU Berlin.

Düker, Heinrch. 1931. *Psychologische Untersuchungen über freie und zwangsläufige Arbeit. Experimentelle Beiträge zur Willens- und Arbeitspsychologie*. Leipzig: Verlag von Johann Ambrosius Barth.

Engeser, Stefan, Falko Rheinberg, and Matthias Möller. 2009. "Achievement Motive Imagery in German Schoolbooks: A Pilot Study Testing McClelland's Hypothesis." *Journal of Research in Personality* 43 (1): 110–14.

Heckhausen, Heinz. 1973. "Einleitung." In *Bedingungen des Bildungsprozesses: Psychologische und pädagogische Forschungen zum Lehren und Lernen in der Schule*, edited by Wolfgang Edelstein and Diether Hopf, 30–9. Stuttgart: Klett.

———. 1976. "Motive und ihre Entstehung." In *Funk-Kolleg. Pädagogische Psychologie*, edited by Franz E. Weinert, Carl Friedrich Graumann, Heinz Heckhausen, and Manfred Hofer, 135–71. Frankfurt am Main: Fischer Taschenbuch.

———. 1980. *Motivation und Handeln. Lehrbuch der Motivationspsychologie*. Berlin, Heidelberg, and New York: Springer-Verlag.

———. 1987. "Vorsatz, Wille und Bedürfnis: Lewin frühes Vermächtnis und ein zugeschütteter Rubikon." In *Jenseits des Rubikon: Der Wille in den Humanwissenschaften*, edited by Peter M. Gollwitzer and Franz E. Weinert, 86–100. Berlin: Springer.

Hentschel, Volker. 1984. "Produktion, Produktivität." In *Geschichtliche Grundbegriffe*, vol. 5, edited by Werner Conze, Otto Brunner, and Reinhart Koselleck, 1–26. Stuttgart: Klett-Cotta.

Herzberg, Frederick. 1959. *Motivation to Work*. New York: Transaction Publishers.

———. 1968. *Work and the Nature of Man*. London: Staples Press.

Hoth, H.M. 1958. *Beitrag zur Klärung des Produktivitätsbegriffes und zur Produktivitätsmessung im Industriebetrieb*. Düsseldorf: Triltsch.

Katona, George. 1962. "The Achieving Society: By David C. McClelland, Princeton: D. Van Nostrand Co., 1961." *The American Economic Review* 52 (3): 580–3.

Knayer, Manfred. 1955. "Analysis and Comparison of Productivity." *Productivity Measurement Review* 2: 4–12.

Knortz, Heike. 2010. "Die Entwicklung des modernen Begriffs von der Arbeitsproduktivität: Ein Beitrag zur Geschichte der deutschen Betriebswirtschaftslehre." *Zeitschrift für Unternehmensgeschichte* 55 (1): 31–51.

Kocherscheid, Gustav. 1975. "Der Mensch als Grenze für 'Wirtschaftlichkeit'." *Rationalisierung* 26 (10): 244–55.

Köster, Roman. 2011. *Die Wissenschaft der Außenseiter. Die Krise der Nationalökonomie in der Weimarer Republik*. Göttingen: Vandenhoeck & Ruprecht.

Kothe, E. 1950. "Die menschliche Grundlage der Refa-Arbeit." *REFA-Nachrichten* 3 (1), 2–4.Kramer, Wolfgang, and Helmut Winter. 1980. *Motivation als zusätzliches Führungsmittel in Klein- und Mittelunternehmen*. Köln: Deutscher Instituts-Verlag.

Mazur, Allan, and Eugen Rosa. 1977. "An Empirical Test of McClelland's 'Achieving Society' Theory'." *Social Forces* 55 (3): 769–74.

McClelland, David. 1961. *The Achieving Society*. New York: Van Nostrand.

———. 1966. *Die Leistungsgesellschaft: Psychologische Analyse der Voraussetzungen wirtschaftlicher Entwicklung*, edited by Ingeborg Y. Wendt and Gerd Fleischmann. Stuttgart: Kohlhammer.

Mierke, Karl. 1955. *Wille und Leistung*. Göttingen: Hogrefe.

Mommsen, Ernst Wolf. 1970. "Wirtschaft und Bildung vor einem Wandel der Werte." *Rationalisierung* 21 (1): 3–6.
Nützenadel, Alexander. 2005. *Stunde der Ökonomen: Wissenschaft, Politik und Expertenkultur in der Bundesrepublik 1949–1974.* Göttingen: Vandenhoeck & Ruprecht.
Oehme, Wolfgang. 1979. *Führen durch Motivation.* Essen: Girardet.
Piketty, Thomas. 2014. *Capital in the Twenty-First Century.* London: The Belknap Press of Harvard University Press.
Productivity Measurement Review. 1955a. "About this Publication." 1: 2–4.
Productivity Measurement Review. 1955b. "Why Measure Productivity?" 1: 2–4.
Productivity Measurement Review. 1962. "Payment for Productivity." 28: 30–4.
Productivity Measurement Review. 1965. "In This Number." 43: 3–6.
Rationalisierung. 1950. "Rationalisierung. Rationalisierung und Leistungssteigerung: Auszug aus dem Memorandum der Deutschen Bundesregierung an die OEEC vom 15.12.1949." 1 (4).
Reuss, Gerhart E. 1960. *Produktivitätsanalyse: Ökonomische Grundlagen und statistische Methodik*, edited by Erwin von Beckerath and Edgar Salin. Basel: Kyklos.
Richman, Raymond L. 1958. "Measuring the Productivity of Office and Administrative Personnel." *Productivity Measurement Review* 14: 21–30.
Rödder, Andreas. 2004. *Die Bundesrepublik Deutschland, 1969–1990.* Munich: Oldenbourg.
Rosenstiel, Lutz von. 1972. *Motivation im Betrieb.* München: Goldmann.
Rubin, Julius. 1961. "The Achieving Society. By David C. McClelland, Princeton: D. Van Nostrand Co., 1961." *The Journal of Economic History* 23 (1): 118–21.
Saldern, Adelheid von, and Rüdiger Hachtmann. 2009. "Das fordistische Jahrhundert: Eine Einleitung." *Zeithistorische Forschungen* 6 (2). http://www.zeithistorische-forschungen.de/16126041-Editorial-2–2009.
Schwake, Elisabeth. 1951. *Der Begriff der Produktivität.* Erlangen: University treatise.
Siegenthaler, Hansjörg. 1993. *Regelvertrauen, Prosperität und Krisen.* Tübingen: J.C.B. Mohr.
Der Spiegel. 1967. "Ballade der Leistung." (April 17).
Uhl, Karsten. 2014. *Humane Rationalisierung. Die Raumordnung der Fabrik im fordistischen Jahrhundert.* Bielefeld: Transcript.
Weigel, H. 1961. "Gauging Manufacturing Efficiency: A Review Article." *Productivity Measurement Review* 25: 55–63.
Weiner, Bernard. 1980. *Human Motivation.* New York: Holt, Rinehart and Winston.
Wiedemann, Albert. 1960. "Die psychischen Komponenten der Leistungsentfaltung." REFA 13(5): 161–4.
Zink, Klaus J. 1975. *Differenzierung der Theorie der Arbeitsmotivation von F. Herzberg zur Gestaltung sozio-technologischer Systeme.* Frankfurt am Main: Verlag Harri Deutsch.

11 Afterword

Histories of Productivity and Modes of Production

Andrew Zimmerman

The essays in this volume offer myriad insights into the variety of ideas and practices of productivity in one particular mode of production: capitalism. Indeed, they suggest that ideas and practices of productivity are themselves constructs of the capitalist mode of production that have become integral components of that mode of production. Taken individually and collectively, these essays elaborate, and add new dimensions to, the classical Marxist understanding of a mode of production as a (contradictory, dialectical) combination of forces of production and relations to production—that is, of the technical processes of production and the broader set of social relations that determine the appropriation and distribution of those products. To capture this rich conception of a mode of production, Peter-Paul Bänziger, Marcel Streng, and Mischa Suter offer the concept of the regime of productivity, made up of "bodies, things/objects, knowledge/concepts, translation/assembly" (this volume, 11). The regime of productivity thus encompasses all elements of Karl Marx's concept of mode of production and incorporates, most importantly, the topics of bodies and knowledge that Marx only hinted at and which Michel Foucault ([1975] 1995; [1976] 1990) did more than anyone else to make available as objects of critical analysis.

Central to many of these essays is the inextricable connection of production, reproduction, and consumption. As Justus Nipperdey suggests in his essay, European thinkers have for centuries discerned a two-way connection between production and reproduction, a relationship of mutual causation between the size of a population and the quantity of goods that it can produce to sustain itself. At the dawn of the capitalist mode of production in Britain, Thomas Malthus, Nipperdey shows, reconfigured this relation, thereby defining a portion of the population as surplus, as an excess whose needs and desires (for food, for sex) threatened the entire social order. These needs and desires, not just for Malthus but for virtually every theorist of capitalism, also underwrote the social order. As the British political economist Joseph Townsend wrote of "the poor", more than a decade before Malthus published his *Essay on the Principle of Population*, "it is only hunger which can spur and goad them on to labour" (Townsend [1786] 1817, 15–16; see

Polanyi [1944] 2001, 119). The valorization of capital encompassed the entirety of workers' lives, even the most basic biological requirements like nutrition. Marx, in places, seemed to image the sphere of reproduction as at least partly separated from the sphere of production, perhaps even offering a kind of respite to the worker: "He is at home when he is not working, and when he is working he is not at home" (Marx 1978, 74). In fact, even if workers could leave the mode of production at their workplace, the regime of productivity, as Monika Streule makes clear in her essay, pursued them across the thresholds of their homes.

The desiring, needy body has remained central to the capitalist mode of production and not just because the body was the object of the "lash of hunger" that liberal humanitarians always imagined would someday obviate the need for actual lashes in the production process. François Jarrige reveals how bakers in France were able to maintain the place of their own bodies—their hands and also, sometimes, their feet—within the production process in France, despite calls for an allegedly more sanitary, mechanical kneading process. Thus, they resisted that transformation, applauded by Andrew Ure (1835, x), in which "every process peculiarly nice, and therefore liable to injury from the ignorance and waywardness of workmen, is withdrawn from handicraft control, and placed under the guidance of self-acting machinery". Ure expressed with unusual frankness widespread authoritarian fantasies of machines replacing bodies in the production process. The machine, for Ure, played the role of Christ for St. Paul, the one "who shall change our vile body, that it may be fashioned like unto his glorious body, according to the working whereby he is able even to subdue all things unto himself" (Philippians, 3: 21). In very rare cases was the victory of machine or body even as clear as in the case of the English textile mills, where machine seemed to triumph over body, or the French bakeries, where body maintained its place in production. Still, the body of the worker remained more or less central to every regime of productivity, simultaneously as a value-creating producer of goods, as a desiring subject of consumption, and as an object of discipline and control. The glorious body of capital, like that of Christ, existed by subduing vile bodies.

Social reform, and especially education, offered one common means of transforming the desiring body so repulsive to Malthus into the body disciplined by its desires so valued by Townsend and others. A number of essays chart how educators, reformers, and business strategists worked on the mind, or on broader cultural forms, to turn the vile body into the productive body. They sought, to borrow a phrase from Foucault ([1975] 1995, 30), to transform "the soul" into "the prison of the body". Encouraging children to abstain from purchasing fireworks and sweets in order to give money instead to the church served, as Sandra Maß relates, not merely as an exercise in generosity but also as a preparation for a productive adult life. The calorie, as Nina Mackert shows, and as many of us experience daily, emerged as a device by which individuals could unceasingly monitor their consumption and bring it into balance with their production. The "moral restraint" that Malthus offered as an alternate to the "misery and vice" to which he had initially condemned humanity could finally be realized in practices identified as healthy. Yet, as Alexa Geisthövel suggests, such pursuits of productive health also depended on a generalization of sickness, as individuals with normal insulin levels were also enjoined to follow "the diabetics' low-fat diet and bodily exercise as a model way of life" (this volume, 149).

In the early twentieth century, Max Weber ([1920] 2001, 123) diagnosed the cultural basis of capitalism as the generalization of the "asceticism" of "monastic cells" into "everyday life", thus confining individuals in an "iron cage" (*stahlhartes Gehäuse*). By the second half of the twentieth century, as Lukas Held shows, social psychologists were building on Weber's grim diagnosis of the capitalist soul so as to devise methods of generalized productivity that they thought should be instilled in workers so that they might better function in the rapidly transforming economic landscape of post-Fordism.

As Cassandra Mark-Thiesen's and Tristan Oestermann's contributions remind us, Europeans were not the only, and not even the majority of, the global working class. There were many echoes between regimes of productivity in Europe and those in regions that Europeans colonized. The stereotypes discussed by Mark-Thiesen not only put forth negative images of lazy Africans but also exempted European capitalists from the more subtle, normalizing forms of power applied to workers defined as fully white. The regimes of productivity that Europeans imposed on their colonies resembled those imposed on racialized or otherwise marginalized others in the capitalist core (Conrad [2006] 2010; Zimmerman 2010). As Achille Mbembe (2003) reminds us, in areas defined as non-white, European power often took the form of murderous necropolitics rather than normalizing biopolitics. Indeed, slavery also continued—and in some areas expanded considerably—within Africa, as well as in the Americas, in the era of emancipation, wage labor, and so-called legitimate trade (Lovejoy 2000). As scholars since Eric Wolf (1982) have highlighted, capitalism was never an exclusively European imposition, but it always incorporated and was even articulated with a variety of locally determined modes, or at least strategies, of production. Oestermann reveals how racist stereotypes authorized violent intervention in Cameroon's rubber economy, but also how African workers engaged in their own economic and political strategies, including class conflicts within African societies.

The regimes of productivity treated in this volume are local, whereas the capitalist mode of production that they each illuminate has long been global, although highly differentiated and constituted at least in part by these differences. As Silvia Federici (2004, 63–4) has written,

> primitive accumulation . . . was not simply an accumulation and concentration of exploitable workers and capital. It was also an accumulation

of differences and divisions within the working class, whereby hierarchies built upon gender, as well as "race" and age, became constitutive of class rule and the formation of the modern proletariat.

There is, in short, no particular worker, no particular regime of productivity, characteristic of, or normative for, the capitalist mode of production. Thus, it is important to theorize not just the various regimes of productivity—or what Jairus Banaji (2011) calls various "modes of exploitation"—but also their combination into a contradictory, highly diverse, and more or less global whole (see Linden 2008).

That no essay in this book directly treats the white, male, wage-earning industrial workers who once, to a certain perspective, appeared as a kind of norm, does not simply redress an imbalance in earlier studies of labor. This volume, rather, frankly acknowledges that there is no capitalist norm for labor—only regimes of productivity applied to bodies, all of them equally vile, desiring, and productive. Marx ([1867] 1976, 128) made clear that the apparent generality of capitalism (a term that, in any case, he did not use) emerges only on the side of capital and only through a "disappearance of the different concrete forms of labor" that can never, in fact, occur. The "glorious body" of capital does not exist: There are only our own vile bodies, in constant and particular struggles against various regimes of productivity. It is as important as ever to understand the capitalist mode of production as a large-scale, even global social formation. However, to understand it primarily as such is to repeat, rather than to historicize, capital's own self-understanding, predicated as it is on "the disappearance of the different concrete forms of labor". As the essays brought together in this volume demonstrate, analyses of regimes of productivity come at capital from every direction but its own, opening multiple sites from which, and multiple analytic paths by which, to grasp the capitalist mode of production.

Bibliography

Banaji, Jairus. 2011. *Theory as History: Essays on Modes of Production and Exploitation*. Chicago: Haymarket Books.

Conrad, Sebastian. (2006) 2010. *Globalisation and the Nation in Imperial Germany*. Translated by Sorcha O'Hagan. Cambridge: Cambridge University Press.

Federici, Silvia. 2004. *Caliban and the Witch: Women, the Body, and Primitive Accumulation*. Brooklyn: Autonomedia.

Foucault, Michel. (1976) 1990. *The History of Sexuality*. Vol 1, translated by Robert Hurley. New York: Vintage.

———. (1975) 1995. *Discipline and Punish: The Birth of the Prison*. Translated by Alan Sheridan, second edition. New York: Vintage Books.

Linden, Marcel van der. 2008. *Workers of the World: Essays Toward a Global Labor History*. Leiden: Brill.

Lovejoy, Paul E. 2000. *Transformations in Slavery: A History of Slavery in Africa*, second edition. Cambridge: Cambridge University Press.

Marx, Karl. (1867) 1976. *Capital: Critique of Political Economy.* Vol 1, translated by Ben Fowkes. New York: Penguin.

———. 1978. "Economic and Philosophic Manuscripts of 1844". In *The Marx-Engels Reader*, second edition, edited by Robert C. Tucker, 66–125. New York: Norton.

Mbembe, Achille. 2003. "Necropolitics." Translated by Libby Meintjes. *Public Culture* 15: 11–40.

Polanyi, Karl. (1944) 2001. *The Great Transformation*, second edition. Boston: Beacon Press.

Townsend, Joseph. (1786) 1817. *A Dissertation on the Poor Laws*. London: Ridgeways.

Ure, Andrew. 1835. *The Philosophy of Manufactures: Or, an Exposition of the Scientific, Moral and Commercial Economy of the Factory System of Great Britain.* London: Charles Knight.

Weber, Max. (1920) 2001. *The Protestant Ethic and the Spirit of Capitalism*, revised edition. Translated by Talcott Parsons. New York: Routledge Classics.

Wolf, Eric R. 1982. *Europe and the People without History*. Berkeley: University of California Press.

Zimmerman, Andrew. 2010. *Alabama in Africa: Booker T. Washington, the German Empire, and the Globalization of the New South*. Princeton: Princeton University Press.

Contributors

Peter-Paul Bänziger is currently the recipient of an Ambizione Grant from the Swiss National Science Foundation at the University of Basel. He studied history and philosophy at the University of Zurich, where he received his PhD in 2008. He held a nontenured assistant professorship at the University of St. Gallen and was a senior scholar at the Center "History of Knowledge" in Zurich. Visiting scholarships led him to the *École des hautes études en sciences sociales* (EHESS) in Paris, the Universities of Vienna and Cologne, and Columbia University in New York City. Presently, he is writing a history of the self in the modern "consumer and work society". His areas of research are body history, media history, and history of knowledge, particularly in the fields of labor and consumption, sexuality, and public health and AIDS. His main publications are *Sex als Problem. Körper und Intimbeziehungen in Briefen an die "Liebe Marta"* (2010) and *Sexuelle Revolution? Zur Geschichte der Sexualität im deutschsprachigen Raum seit den 1960er Jahren* (edited 2015, with M. Beljan, F. X. Eder, and P. Eitler).

Alexa Geisthövel is a member of the ERC-funded research group "Ways of Writing: How Physicians Know, 1550–1950" at the Institute for the History of Medicine in Berlin. As a historian of the nineteenth and twentieth centuries, she has worked on a variety of subjects, including the history of anthropology and intelligence testing in Germany and the US; these efforts were published in the monograph *Intelligenz und Rasse. Franz Boas' psychologischer Antirassismus zwischen Amerika und Deutschland, 1920–1942* (2013).

Lukas Held is a PhD student at the Research Institute for Social History and the History of Economics at the University of Zurich. His work focuses on the constitution of motivation psychology as a subdiscipline in West Germany since the 1950s and on concepts of achievement. He studied contemporary history and literature at Albert Ludwigs University in Freiburg, Humboldt University in Berlin and the University of Leicester. After graduating in 2011, he worked as a research assistant in Veronika Lipphardt's research group at the Max Planck Institute for the History of Science before moving to Zurich in July 2013. His research interests lie in the cultural history of the Cold War; the history of divided Germany, with a focus on sports, schooling, and work life; and the history of psychology, with regard to achievement motivation, exhaustion, and practices of measurement.

François Jarrige is an associate professor in contemporary history at the Université of Burgundy (Dijon), where he is a member of the Centre Georges Chevrier research unit. He works on the history of industrial society and investigates the conflicts, debates, and controversies that have accompanied technological change and industrialization in the West. Among his publications are *Les luddites: Bris de machines, économie politique et histoire* (2006, coauthored with V. Bourdeau and J. Vincent), *Au temps des "tueuses de bras": Les bris de machines à l'aube de l'ère industrielle (1780–1860)* (2009), and *Technocritiques. Du refus des machines à la contestation des technosciences* (2014).

Nina Mackert is an assistant professor of North American history at the University of Erfurt. Currently, she is working on a book on the history of the calorie as part of the joint research project "Food, Health, and Modern Societies: Germany and the United States". She obtained her PhD in history in 2012 from the University of Erfurt with a dissertation on the productivity of the US-American delinquency scare between the 1940s and 1960s (published in 2014 with Universitätsverlag Konstanz/UVK). Recent publications include works on the history of food, body fat, and subjectivation in the decades around 1900.

Cassandra Mark-Thiesen is a lecturer and researcher in the History Department at the University of Basel. Her research focus lies in the social and economic history of West Africa and global labor history. She is working on a book to be published by the Rochester University Press on the recruitment of wage earners in colonial Ghana's first mechanized gold mines through West African intermediaries, the headmen. General research interests include Africa in the nineteenth and twentieth centuries, the age of developmentalism (decolonization), rural history, the history of African intermediaries, and the history of work.

Sandra Maß is deputy director of the Georg Eckert Institute/Leibniz Institute for International Textbook Research. She has published on the history of colonialism and masculinity, development aid, and the cultural history of capitalism. Among her publications are the monographs *Weiße Helden, schwarze Krieger. Zur Geschichte kolonialer Männlichkeit in Deutschland, 1918–1964* (2006) and *Kinderstube des Kapitalismus. Monetäre Erziehung im 18. und 19. Jahrhundert* (2016, forthcoming). Together with Xenia von Tippelskirch, she edited *Faltenwürfe der Geschichte. Entdecken, entziffern, erzählen* (2014). A special issue of *L'Homme. Europäische Zeitschrift für Geschichtswissenschaft* on

"Ökonomien" will be published by Margareth Lanzinger, Claudia Opitz, and Sandra Maß in 2016.

Justus Nipperdey studied modern history and economics at the University of Bonn, the University of Caen, and the University of Munich. He received his doctoral degree in 2009, having studied population thought and population policy in early modern Germany. Since 2010, he is a Wissenschaftlicher assistant/assistant professor at Saarland University, working on the history of the concept of "early modern" in twentieth-century historiography. His research focuses on the history of economic thought, the history of population politics and migration in early modern Europe, the history of historiography, and the organization of the historical profession in the twentieth century. His main publications are *Die Erfindung der Bevölkerungspolitik. Staat, politische Theorie und Population in der Frühen Neuzeit* (2012), "Die Hugenottenaufnahme als Katalysator der Idee des Populationismus" (2013, in: *Francia*), "Johann Peter Süßmilch: From Divine Law to Human Intervention" (2011, in: *Population (E)*).

Tristan Oestermann is a PhD student in history at Humboldt University in Berlin. His dissertation project examines the labor history of the rubber trade in the German colony of Cameroon, and an Elsa Neumann Scholarship funds the work. His research interests focus primarily on the history of African-European contact and interaction in West and Central Africa, from colonial trade and resource exploitation to development policy and humanitarian aid.

Marcel Streng studied modern history, French linguistics, sociology, and public law at the University of Marburg, Humboldt University in Berlin, and Toulouse/Le Mirail University in France. He has worked as a research associate at the University of Bielefeld and the University of Cologne, as well as the Ludwig Boltzmann Institute for European History and Public Spheres in Vienna. His PhD project on local food market policies in France (1846–1914) was funded by the Deutsche Forschungsgemeinschaft and completed at the Research Center on Political History at the University of Bielefeld in 2014. He is currently working on the body history of hunger in nineteenth and twentieth century Europe and social therapy, prison policies, and the inmate's self in West Germany during the 1970s. Primary publications related to the topic of the volume are " 'Vie chère', Violent Protest, and Visions of Protection: The Vie chère Controversy in France on the Eve of WWI (1905–1913)" (2012, in: *Writing Political History Today*, edited, by W. Steinmetz, I. Holtey, and H. G. Haupt) and "The Food Riot Revisited: New Dimensions in the History of 'Contentious Food Politics' in Germany before WWI" (2013, in: *European Review of History*).

Monika Streule is a lecturer and researcher with the Architecture Faculty of ETH Zürich. As an urban anthropologist, she has worked on a variety of subjects, including urban transformation in Mexico City, focusing particularly on questions regarding the constitution and everyday practices of appropriation of urban space. Her current research focuses on urbanization from a historical, ethnographic, and comparative perspective in the context of her PhD thesis and in the research project "Urbanization in Comparative Perspective" at the Future Cities Laboratory (Singapore-ETH Centre for Global Environmental Sustainability). Visiting scholarships led her to the Instituto de Investigaciones Antropológicas (IIA) in 2013, and to the Instituto de Investigaciones Sociales (ISS) in 2005, both at the Autonomous National University of Mexico (UNAM). Among her publication are "La Festivalización de los Centros Históricos" (2008, in: *Ciudades*), "The Trend Towards Transdisciplinarity: A Critical Evaluation of an Ambivalent Practice in Qualitative Urban Studies" (2014, in: *Forum Qualitative Sozialforschung/Forum: Qualitative Social Research*), and her PhD thesis *Patrones y Trayectorias de una Metropoli. Eine transdisziplinäre Analyse aktueller Urbanisierungsprozesse von Mexiko-Stadt* (2016, forthcoming).

Mischa Suter is a lecturer in Modern European history at the University of Basel. He studies the history of everyday life under nineteenth-century capitalism. His interests include the historical epistemology of political economy, the history of legal practices, and the making of the social. His first book deals with debt collection in liberal capitalism and is forthcoming with Konstanz University Press.

Andrew Zimmerman is professor of history at George Washington University (Washington, DC). He is the author of *Anthropology and Antihumanism in Imperial Germany* (2001) and *Alabama in Africa: Booker T. Washington, the German Empire, and the Globalization of the New South* (2010). He is also the editor of *Karl Marx and Friedrich Engels, The Civil War in the United States* (2016, forthcoming). He is currently writing a history of the American Civil War as a transnational revolution against slave labor and wage labor.

Index

Numbers in bold indicate terms appearing in chapter titles.

For Product Safety Concerns and Information please contact our EU
representative GPSR@taylorandfrancis.com
Taylor & Francis Verlag GmbH, Kaufingerstraße 24, 80331 München, Germany

www.ingramcontent.com/pod-product-compliance
Ingram Content Group UK Ltd.
Pitfield, Milton Keynes, MK11 3LW, UK
UKHW021720240926
475702UK00029BA/1537

* 9 7 8 1 1 3 8 3 2 9 5 1 5 *